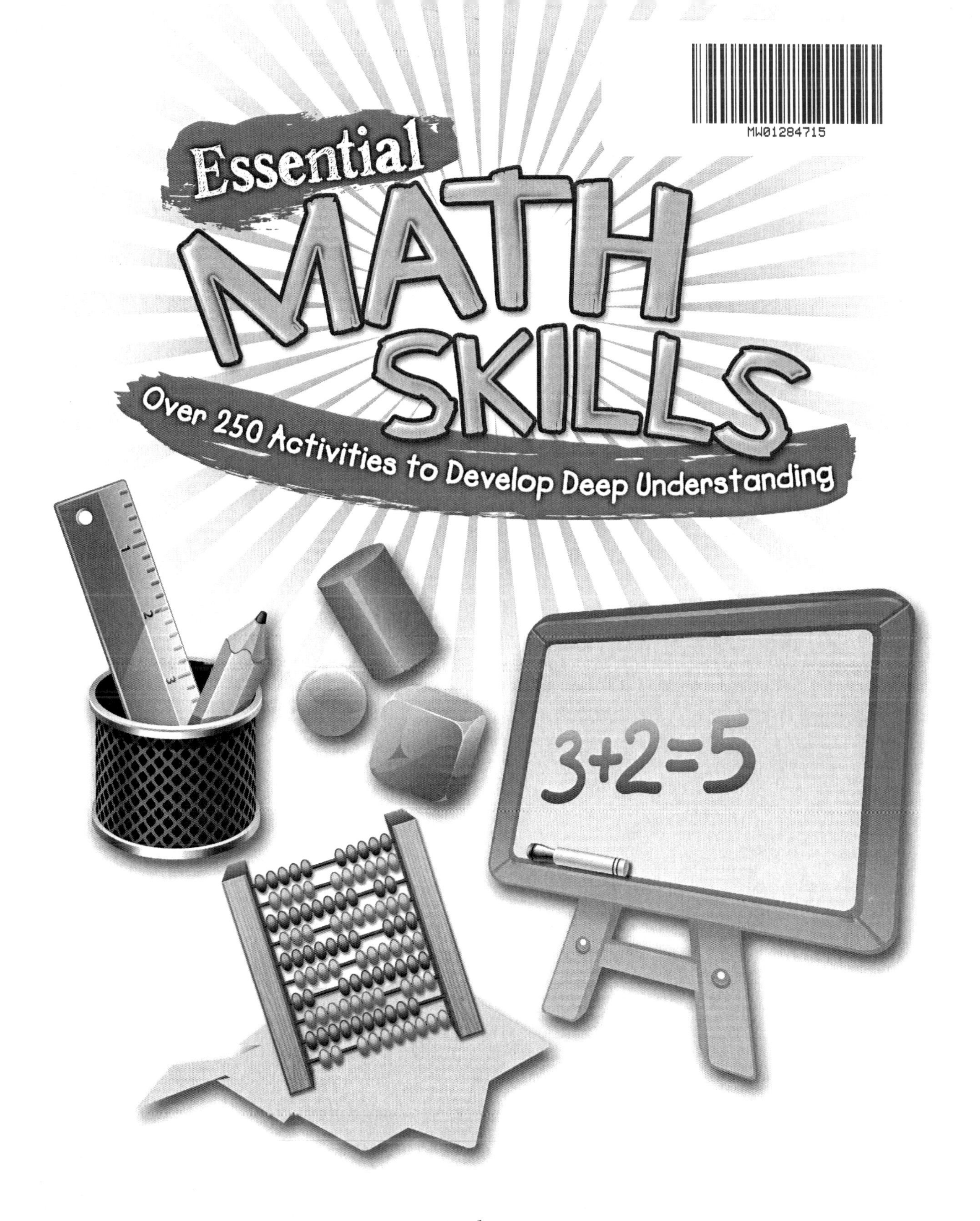

Author
Bob Sornson, Ph.D.

Publishing Credits

Robin Erickson, *Production Director*; Lee Aucoin, *Creative Director*;
Timothy J. Bradley, *Illustration Manager*; Sara Johnson, M.S.Ed., *Editorial Director*;
Aubrie Nielsen, M.S.Ed., *Senior Editor*; Grace Alba Le, *Designer*;
Corinne Burton, M.A.Ed., *Publisher*

Image Credits

p.48 swanyland/iStockphoto; p.49, 89 azureforest/iStockphoto; p.67, 101 FrankRamspott/iStockphoto;
All other images Shutterstock.

Standards

Essential Math Skills Inventories and Rubrics

Shell Education
5301 Oceanus Drive
Huntington Beach, CA 92649-1030
http://www.shelleducation.com

ISBN 978-1-4258-1211-9

Table of Contents

Research

Some math skills are simply essential to understanding numbers and how they work. These are the skills that might be considered the "core of the Core." They must be well understood, or a student will be forever compromised as he or she moves forward through more complex mathematics.

When learning these foundational math skills, racing through content is not a suitable plan for instruction. For these essential skills, "drill and kill" is not acceptable. The foundation for loving math for life must be built on success and joyful learning.

> *U.S. curricula typically include many topics at each grade level, with each receiving relatively limited development, while top-performing countries present fewer topics at each grade level but in greater depth. In addition, U.S. curricula generally review and extend at successive grade levels many (if not most) topics already presented at earlier grade levels, while the top-performing countries are more likely to expect closure after exposure, development, and refinement of a particular topic.*
>
> — *National Mathematics Advisory Panel* (20, 2008)

For decades, our schools have been engaged in a failed experiment that has attempted to cram more content into the time available for instruction than is humanly possible. Most schools have asked students to learn overwhelmingly complex content at younger and younger ages without carefully building the foundational skills needed for learning success.

Early childhood is the crucial time during which one builds the foundational skills, behaviors, and beliefs that establish one's path as a learner for life. Sadly, for many young students, our teaching systems are not working effectively. By the beginning of fourth grade, the point at which we can accurately predict long-term learning outcomes, only 40 percent of American students perform at proficient levels in mathematics. By eighth grade, this has decreased to 35 percent student proficiency, and by twelfth grade, only 26 percent of students remaining in school perform at or above a proficient level in mathematics (National Assessment of Educational Progress 2009 and 2011).

The long-term effects of such numbers of American students lacking proficiency in mathematics in the information age may be disastrous. Low-skill learners become low-skill workers with low wages. Early learning success in reading and mathematics is correlated with high school graduation, going on to advanced education, lower rates of substance abuse and other high-risk behaviors, decreased criminality, stable adult relationships, and success on the job (Hernandez 2011; Shore 2009a; Shore 2009b; The Annie E. Casey Foundation 2010). The costs of allowing three-quarters of our students to remain nonproficient in math include diminished employment options for our students and reduced prosperity for our society.

Research *(cont.)*

With the belief that many more math students could deeply understand basic math skills, become successful math students, and enjoy using mathematics in their lives, the Early Learning Foundation set out to identify those nonnegotiable skills that need to be deeply understood in order for a student to have continued success in math learning. The foundation began with backward planning; as an example, researchers asked second-grade teachers which first-grade skills were absolutely needed for success in second grade. Instead of long lists of content standards to "cover," these teachers identified the core skills that provide learning readiness for ongoing mathematics instruction.

The sequential nature of learning in mathematics provides a beautiful framework for what is essential. The skills identified by teachers matched the accepted theoretic framework for the development of mathematical learning. By experimenting with the exact wording of the skills and the rubrics in several schools in three districts, over a three-year period, researchers refined an inventory of essential skills for success in mathematics and a corresponding rubric to assess students' mastery of those skills. With more years of experience in additional schools, the rubric has been further refined so that it can be implemented with any high-quality mathematics curriculum.

By identifying crucial skills and encouraging teachers to take all the time needed to help students develop these skills, teachers began to observe that incoming students were better prepared than ever before, and scores on state or standardized tests increased dramatically. Equally important, more students enjoyed math and saw themselves as successful math learners.

The goals of using the Essential Math Skills Inventories are:

- Identify the essential math skills in sequence and at each level.
- Provide a concise set of outcomes that can be systematically tracked using teacher observation and ongoing formative assessment.
- Help teachers remember that students learn best while learning in the instructional zone, which includes a small element of challenge along with the experience of understanding and success.
- Give teachers data to reinforce the need to allow students sufficient time to learn a skill.
- Provide a structure to help teachers verify proficiency so that they can know how to plan differentiated instruction in mathematics based on students' individual readiness levels and when to move them along to more challenging math learning activities.

Every school district has an instructional framework for what should be covered at each grade level. But essential skills deserve more than coverage. They must be learned to a level of deep understanding and application. They are the core of the Common Core. They provide a framework for formative assessment, allowing teachers to monitor progress toward those crucial outcomes. These skills are the solid foundation we want for every child.

The Essential Math Skills for Grades Pre-K–3

Pre-Kindergarten Skills	
Skill 1: Demonstrates one-to-one correspondence for numbers 1–10, with steps **Skill 2:** Demonstrates one-to-one correspondence for numbers 1–10, with manipulatives	**Skill 3:** Adds on, using numbers 1–10, with steps **Skill 4:** Adds on, using numbers 1–10, with manipulatives
Kindergarten Skills	
Skill 5: Demonstrates counting to 100 **Skill 6:** Has one-to-one correspondence for numbers 1–30	**Skill 7:** Understands combinations (within 10) **Skill 8:** Recognizes number groups without counting (2–10)
Grade 1 Skills	
Skill 9: Counts objects with accuracy to 100 **Skill 10:** Replicates visual or movement patterns **Skill 11:** Understands concepts of adding on or taking away (within 30), with manipulatives	**Skill 12:** Adds/subtracts single-digit numbers on paper **Skill 13:** Shows a group of objects by number (to 100)
Grade 2 Skills	
Skill 14: Quickly recognizes groups of objects (to 100) **Skill 15:** Adds to/subtracts from a group of objects (within 100) **Skill 16:** Adds/subtracts two-digit numbers on paper	**Skill 17:** Counts by 2, 3, 4, 5, and 10, using manipulatives **Skill 18:** Solves written and oral story problems using the correct operations *(addition and subtraction)* **Skill 19:** Understands/identifies place value to 1,000
Grade 3 Skills	
Skill 20: Reads and writes numbers to 10,000 in words and numerals **Skill 21:** Uses common units of measurement: length, weight, time, money, and temperature **Skill 22:** Adds/subtracts three-digit numbers on paper with regrouping **Skill 23:** Rounds numbers to the nearest 10 **Skill 24:** Rounds numbers to the nearest 100	**Skill 25:** Adds/subtracts two-digit numbers mentally **Skill 26:** Counts by 5, 6, 7, 8, 9, and 10 using manipulatives **Skill 27:** Uses arrays to visually represent multiplication **Skill 28:** Recognizes basic fractions **Skill 29:** Solves written and oral story problems using the correct operation *(addition, subtraction, and grouping)*

Implementing *Essential Math Skills* in the Classroom

Essential Math Skills is designed to help students develop the numeracy skills that will allow them to understand math, love math, and succeed in the information economy. The activities in this book provide a rich menu of mathematical learning experiences that include the use of manipulatives, exploration, inquiry, and play.

Essential Math Skills allows teachers to:

- Understand the sequence of essential early math skills
- Identify the essential early math skills by grade level
- Determine which skills each student has mastered
- Design responsive instruction to meet the needs of each student
- Understand which skills require additional instructional experiences
- Systematically monitor progress toward proficiency in every essential skill
- Advance students as soon as they are ready for more complex instruction
- Help each student develop a solid foundation of mathematical skills and concepts
- Help many more students fall in love with math for life

Using *Essential Math Skills* is simple. The *Skill Progression Rubrics* (see Appendix A) will help the teacher identify the skills that each student has already developed to proficiency, and then the teacher can plan instruction based on the learning readiness of the students. Skill-specific activities offer many different options for engaging students in learning that is challenging but never overwhelming.

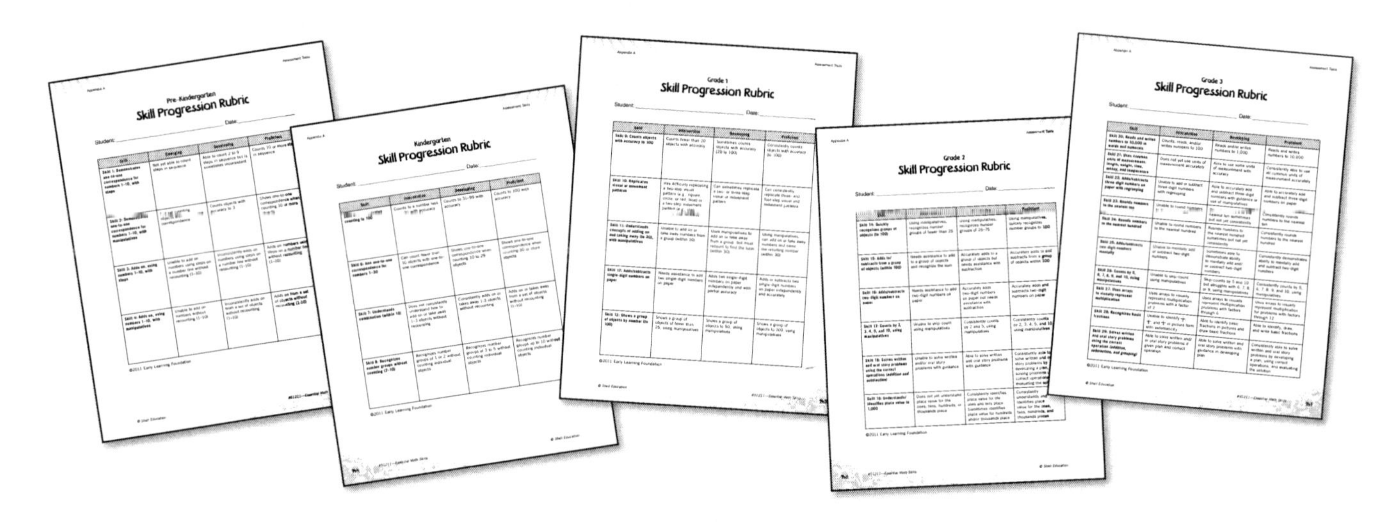

Implementing *Essential Math Skills* in the Classroom *(cont.)*

Using the *Essential Math Skills* Inventories

The *Essential Math Skills Class Inventories* (pages 137–141; see digital resources on page 168) and the *Essential Math Skills Individual Inventory* (pages 135–136; individualinventory.pdf; individualinventory.xls) offer simple formats for systematically assessing the 29 most crucial skills in the development of early mathematical literacy. The inventories serve as ongoing formative assessment tools for identifying specifically what students know and what they are ready to learn. These skills are at the core of mathematical literacy and cannot be merely "covered." They are the skills students must learn to a level of deep understanding and application, and the foundation upon which a lifetime of successful mathematical learning will be built.

Classroom teachers will use the *Essential Math Skills Class Inventory* as their primary math-data collection tool. This will give teachers all the information needed to understand the learning readiness and needs of every student in the class on one page, allowing for well-designed instruction that meets the specific needs of the students. Classroom teachers can use the *Individual Essential Math Skills Inventory* to communicate with parents about the learning progress of individual students. Teachers are advised not to show class inventory data to parents since this would compromise the privacy of other students and could create unnecessary competition. When working with students with a wide range of skills, special education or intervention instructors may wish to use the *Individual Essential Math Skills Inventory* as their primary math-data collection tool. This would require monitoring more inventories but would also allow those teachers to accurately track the varied learning needs and progress of individual students.

1. During the first few weeks of school, **use observational, informal, and instructional assessment to collect baseline data**. Using the *Skill Progression Rubric* for your grade level (Appendix A), determine which grade-level skills each student has already developed proficiency in and which may need intensive support. Determine which students may need to spend time further developing a skill from a previous grade. **Note:** *The* Essential Math Skills Class Inventory *allows the teacher to monitor progress on skills that are one grade level below and one grade level above the current grade.*

2. **Use the *Proficiency Checklist*** (page 142) to keep track of students' progress toward proficiency in a skill.

3. **Update the class inventory each week**. Systematic formative assessment is necessary for responsive instruction. Note when proficiency is achieved for each student by recording the date on the inventory. Student proficiency should be documented only after the student has demonstrated **proficiency for a skill on at least three occasions, using more than one type of instructional activity**. Be certain that a student deeply understands a skill and can apply it in multiple scenarios before confirming proficiency.

Implementing *Essential Math Skills* in the Classroom *(cont.)*

4. **Use individual student data to communicate with parents.** The individual inventory allows teachers to evaluate students' performance on skills as *Emerging/Intervention*, *Developing*, or *Proficient*. Teachers can ask parents to support the development of crucial skills at home. Suggestions for ways parents can do this can be found in the *Family Letter* (page 165).

5. **Use the data to plan instruction.** Refer to this information as you formulate weekly or daily lesson plans. With the knowledge of which students are proficient, developing, or need intensive support in certain skills, teachers can design small-group lessons, centers, and activities to meet the specific needs of the students.

6. **Work to help students reach or exceed proficiency in every grade-level skill** by the end of the school year. Students will undoubtedly move through the skills at varying rates; some will need many more experiences with a skill to reach proficiency than will others. Many students will exceed grade level standards and be ready to move on to more advanced skills on the inventory.

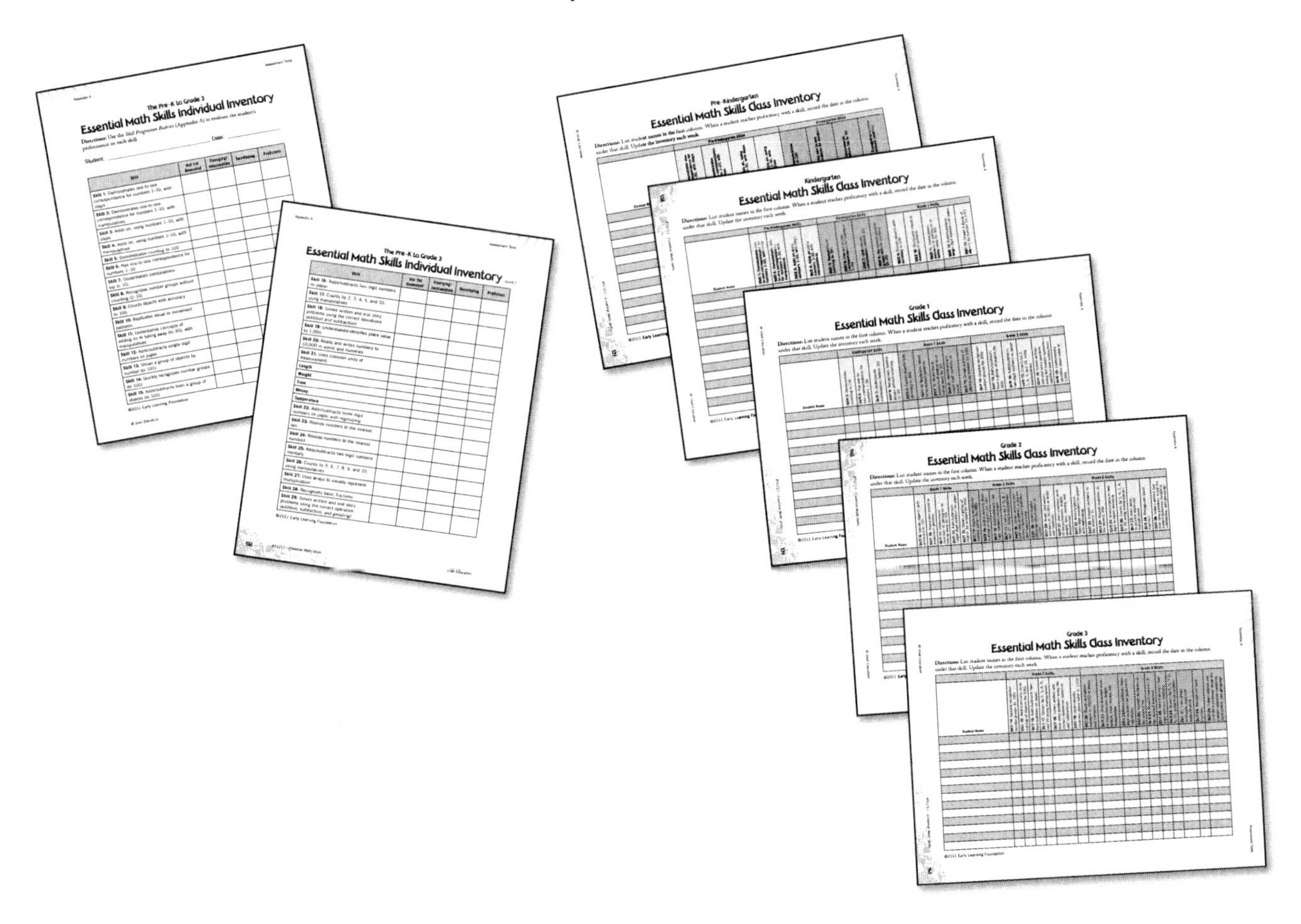

How to Use This Book

Skills and Activities Overview

The **divider page** at the beginning of each grade-level section offers an overview of the skills and materials needed for the activities.

The **Recommended Materials** lists all resources that are necessary to lead students in the activities for the grade level. **Note:** *Several activities within this resource use a 10-by-10 counting frame, usually called an* abacus. *This tool is used to learn the value of numbers using a base-ten format and to support a deep understanding of basic computation. The abacus allows students to understand the feeling of moving a number value, see base-ten displays of number values, and see the relationship between number values used in addition, subtraction, and grouping. Some teachers may remember the days when every early childhood classroom had a large classroom abacus. For others, this will be a new and exciting tool that allows students have a kinesthetic, visual, and auditory experience to help them deeply understand basic mathematical processes.*

The **Essential Math Skills** lists the focus skills for each grade level. These are the core skills with which students must be completely proficient so that they can understand and enjoy mathematical learning.

The **Additional Resources** lists activity sheets that are provided in the appendix or on the Digital Resource CD and that support the activities for the grade level.

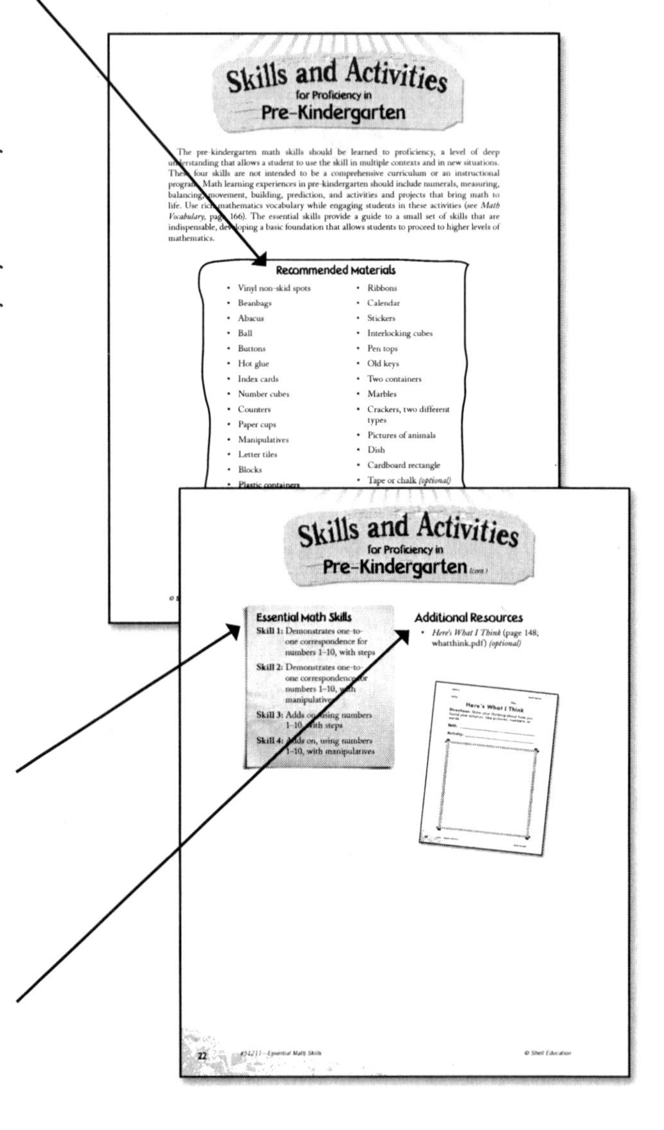

Skills and Activities for Proficiency in Pre-Kindergarten

The pre-kindergarten math skills should be learned to proficiency, a level of deep understanding that allows a student to use the skill in multiple contexts and in new situations. These four skills are not intended to be a comprehensive curriculum or an instructional program. Math learning experiences in pre-kindergarten should include numerals, measuring, balancing, movement, building, prediction, and activities and projects that bring math to life. Use rich mathematics vocabulary while engaging students in these activities (see *Math Vocabulary*, page 166). The essential skills provide a guide to a small set of skills that are indispensable, developing a basic foundation that allows students to proceed to higher levels of mathematics.

Recommended Materials

- Vinyl non-skid spots
- Beanbags
- Abacus
- Ball
- Buttons
- Hot glue
- Index cards
- Number cubes
- Counters
- Paper cups
- Manipulatives
- Letter tiles
- Blocks
- Plastic containers
- Ribbons
- Calendar
- Stickers
- Interlocking cubes
- Pen tops
- Old keys
- Two containers
- Marbles
- Crackers, two different types
- Pictures of animals
- Dish
- Cardboard rectangle
- Tape or chalk *(optional)*

Skills and Activities for Proficiency in Pre-Kindergarten (cont.)

Essential Math Skills

Skill 1: Demonstrates one-to-one correspondence for numbers 1–10, with steps

Skill 2: Demonstrates one-to-one correspondence for numbers 1–10, with manipulatives

Skill 3: Adds on, using numbers 1–10, with steps

Skill 4: Adds on, using numbers 1–10, with manipulatives

Additional Resources

- *Here's What I Think* (page 148; whatthink.pdf) *(optional)*

22 #51211—Essential Math Skills © Shell Education

How to Use This Book *(cont.)*

Skills and Activities Overview

The **Skill Progression** defines proficiency for each skill and provides a concise rubric for evaluating students' progress toward mastery.

The **Activities for Skill Development and Assessment** section offers detailed descriptions of the activities and experiences that will enable students to reach proficiency in a skill.

Variations offer creative modifications that allow activities to be revisited for additional practice.

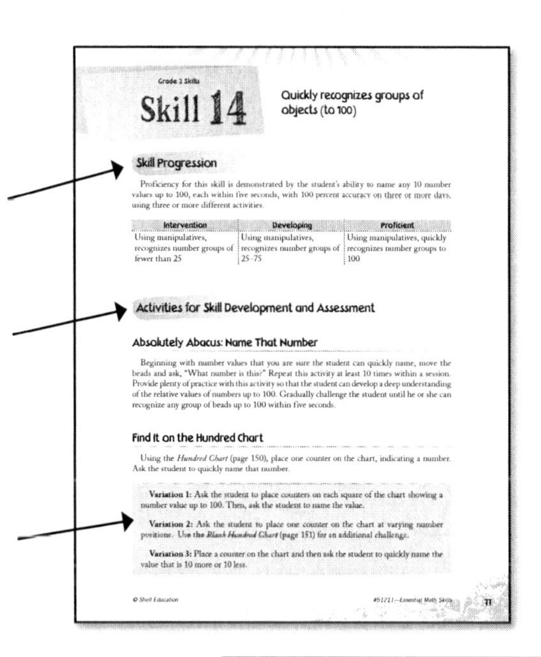

Grade 2 Skills

Skill 14

Quickly recognizes groups of objects (to 100)

Skill Progression

Proficiency for this skill is demonstrated by the student's ability to name any 10 number values up to 100, each within five seconds, with 100 percent accuracy on three or more days, using three or more different activities.

Intervention	Developing	Proficient
Using manipulatives, recognizes number groups of fewer than 25	Using manipulatives, recognizes number groups of 25–75	Using manipulatives, quickly recognizes number groups to 100

Activities for Skill Development and Assessment

Absolutely Abacus: Name That Number

Beginning with number values that you are sure the student can quickly name, move the beads and ask, "What number is this?" Repeat this activity at least 10 times within a session. Provide plenty of practice with this activity so that the student can develop a deep understanding of the relative values of numbers up to 100. Gradually challenge the student until he or she can recognize any group of beads up to 100 within five seconds.

Find It on the Hundred Chart

Using the *Hundred Chart* (page 150), place one counter on the chart, indicating a number. Ask the student to quickly name that number.

Variation 1: Ask the student to place counters on each square of the chart showing a number value up to 100. Then, ask the student to name the value.

Variation 2: Ask the student to place one counter on the chart at varying number positions. Use the *Blank Hundred Chart* (page 151) for an additional challenge.

Variation 3: Place a counter on the chart and then ask the student to quickly name the value that is 10 more or 10 less.

© Shell Education #51211—Essential Math Skills 11

Activities to Extend Deep Understanding offer ideas to extend the skill after students have shown proficiency through the Activities for Skill Development and Assessment.

The **Digital Resource CD** offers resources such as student activity sheets and data-recording tools.

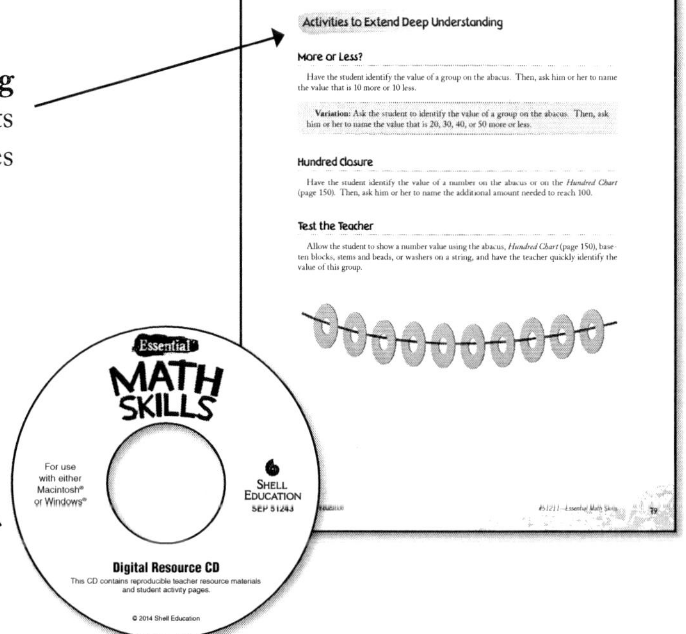

Skill 14

Activities to Extend Deep Understanding

More or Less?

Have the student identify the value of a group on the abacus. Then, ask him or her to name the value that is 10 more or 10 less.

Variation: Ask the student to identify the value of a group on the abacus. Then, ask him or her to name the value that is 20, 30, 40, or 50 more or less.

Hundred Closure

Have the student identify the value of a number on the abacus or on the *Hundred Chart* (page 150). Then, ask him or her to name the additional amount needed to reach 100.

Test the Teacher

Allow the student to show a number value using the abacus, *Hundred Chart* (page 150), base-ten blocks, stems and beads, or washers on a string, and have the teacher quickly identify the value of this group.

#51211—Essential Math Skills 19

Tips for Planning Instruction and Assessment in Mathematics

Let's help the next generation of students fall in love with math. Learning about numbers and mathematical thinking should be about exploration and inquiry, and patterns and play. Consider the following recommendations when planning mathematics instruction for students in the early grades.

- **Plan instructional activities and projects that the student can do with a high rate of success.** Students work harder and longer and stay far more engaged when they are good at a task. Generally, students should experience success at least 90 to 95 percent of the time to stay in the optimal state of mind for maximum effort and learning. Don't push too fast and cause students to move out of this optimal learning state.
- **Include movement and manipulatives in your instruction.** Students need concrete learning experiences, including those that involve lifting, stepping, carrying, touching, hearing, and seeing. As students advance, plan instruction that allows them to see the solution to a problem with manipulatives (e.g., an abacus, a ten-frame, beads, or blocks) and then write the solution to the problem on paper. Students need to understand mathematical concepts at the concrete level before they will truly understand them at the symbolic level.
- **Encourage students to justify their thinking and reasoning.** Ask students to explain the choices they make while problem solving. Encourage them to share the reasoning behind their solution paths. Have students justify their responses whether their answers are correct or incorrect. Use the *Here's What I Think* activity sheet (page 148) with any activity when you want students to record their reasoning on paper.
- **Carefully monitor each student for engagement, interest, success, and frustration.** This is called *observational assessment* or *formative assessment.* Use your observations to help you continue or adjust your instruction. Give each student an opportunity to engage in essential skills activities each day.
- **Plan projects and activities, not just rote practice of basic math facts or endless activity sheets.** Projects help make math come alive, offering an opportunity to see how math concepts connect to building, digging, drawing, deconstructing, imagining, planning, and working together as a team.
- **Use *Essential Math Skills* in combination with other instructional tools** and materials to plan a standards-based curriculum.
- **Develop clear routines for the delivery of mathematics instruction and learning time.** Routines for getting out materials, working together, working independently, checking work, and putting away materials all add to the predictable environment that supports good learning.

Tips for Planning Instruction and Assessment in Mathematics *(cont.)*

- **Take your time.** The standard for proficiency used in *Essential Math Skills* is the acquisition of deep understanding of the concept or skill as well as the ability to use the skill easily and in many different learning contexts. Learning these skills should be neither a race nor a contest for students. The importance of these skills demands that teachers accept students where they are, provide them with all the time, encouragement, and practice necessary to learn the skills, and systematically monitor their progress in order to move them along to the next level of challenge as soon as they are fully ready.
- **Stand up for the students and speak up for best practice.** In many schools, teachers have been asked to "cover" more math standards than are humanly possible, treat all kids as if they are at the same level of math skills, and "keep up" with the pacing guide. An educational model that encourages a race for grades and that sorts students into winners and losers during the crucial early learning years, is neither effective nor ethical. We can help almost every student develop a deep understanding of the essential math skills.

With less emphasis on racing through content, teachers can identify essential learning outcomes and use ongoing formative assessment to keep track of how each student is progressing toward the skills that matter most. Teachers can help students build every foundational skill to a proficient level, help more students love math, and bring more joy back into our classrooms.

Teach math with love. Teach students with love. Help students fall in love with math!

Correlation to Standards

Shell Education is committed to producing educational materials that are research- and standards-based. In this effort, we have correlated all of our products to the academic standards of all 50 United States, the District of Columbia, the Department of Defense Dependents Schools, and all Canadian provinces.

How to Find Standards Correlations

To print a customized correlation report of this product for your state, visit our website at http://www.shelleducation.com and follow the on-screen directions. If you require assistance in printing correlation reports, please contact Customer Service at 1-877-777-3450.

Purpose and Intent of Standards

Legislation mandates that all states adopt academic standards that identify the skills students will learn in kindergarten through grade twelve. Many states also have standards for Pre-K. This same legislation sets requirements to ensure the standards are detailed and comprehensive.

Standards are designed to focus instruction and guide adoption of curricula. Standards are statements that describe the criteria necessary for students to meet specific academic goals. They define the knowledge, skills, and content students should acquire at each level. Standards are also used to develop standardized tests to evaluate students' academic progress. Teachers are required to demonstrate how their lessons meet state standards. State standards are used in the development of all of our products, so educators can be assured they meet the academic requirements of each state.

Common Core State Standards

Many lessons in this book are aligned to the Common Core State Standards (CCSS). The standards support the objectives presented throughout the lessons and are provided on the Digital Resource CD (standards.pdf).

TESOL and WIDA Standards

The lessons in this book promote English language development for English language learners. The standards listed on the Digital Resource CD (standards.pdf) support the language objectives presented throughout the lessons.

Correlation to Standards *(cont.)*

Common Core State Standards

The Essential Math Skills represent the skills that must be learned to a level of deep understanding and application. The Common Core State Standards establish standards for instruction and identify content that should be included in the mathematics curriculum for each grade level. The Essential Math Skills are skills with which proficiency is required to meet those standards.

Pre-Kindergarten* and Kindergarten	
Standard	**Skill**
K.CC.1 Count to 100 by ones and by tens.	Skill 1 (pages 23–26); Skill 2 (pages 27–30); Skill 3 (pages 31–33); Skill 5 (pages 39–43); Skill 6 (pages 44–46); Skill 7 (pages 47–50)
K.CC.2 Count forward beginning from a given number within the known sequence (instead of having to begin at 1).	Skill 3 (pages 31–33); Skill 4 (pages 34–36); Skill 7 (pages 47–50)
K.CC.4 Understand the relationship between numbers and quantities; connect counting to cardinality.	Skill 1 (pages 23–26); Skill 2 (pages 27–30); Skill 3 (pages 31–33); Skill 4 (pages 34–36); Skill 6 (pages 44–46); Skill 7 (pages 47–50); Skill 8 (pages 51–53)
K.CC.5 Count to answer "how many?" questions about as many as 20 things arranged in a line, a rectangular array, or a circle, or as many as 10 things in a scattered configuration; given a number from 1–20, count out that many objects.	Skill 1 (pages 23–26); Skill 2 (pages 27–30); Skill 3 (pages 31–33); Skill 4 (pages 34–36); Skill 6 (pages 44–46); Skill 7 (pages 47–50); Skill 8 (pages 51–53)
K.CC.6 Identify whether the number of objects in one group is greater than, less than, or equal to the number of objects in another group (e.g., by using matching and counting strategies).	Skill 3 (pages 31–33); Skill 4 (pages 34–36); Skill 7 (pages 47–50)
K.OA.1 Represent addition and subtraction with objects, fingers, mental images, drawings, sounds (e.g., claps), acting out situations, verbal explanations, expressions, or equations.	Skill 3 (pages 31–33); Skill 4 (pages 34–36); Skill 7 (pages 47–50)

*The pre-kindergarten skills are aligned to those kindergarten standards for which they begin to build understanding.

Correlation to Standards *(cont.)*

Grade 1	
Standard	**Skill**
1.OA.1 Use addition and subtraction within 20 to solve word problems involving situations of adding to, taking from, putting together, taking apart, and comparing, with unknowns in all positions (e.g., by using objects, drawings, and equations with a symbol for the unknown number to represent the problem).	Skill 11 (pages 64–67); Skill 12 (pages 68–70)
1.OA.2 Solve word problems that call for addition of three whole numbers whose sum is less than or equal to 20 (e.g., by using objects, drawings, and equations with a symbol for the unknown number to represent the problem).	Skill 11 (pages 64–67); Skill 12 (pages 68–70)
1.OA.3 Apply properties of operations as strategies to add and subtract.	Skill 12 (pages 68–70)
1.OA.4 Understand subtraction as an unknown-addend problem.	Skill 11 (pages 64–67); Skill 12 (pages 68–70)
1.OA.5 Relate counting to addition and subtraction (e.g., by counting on 2 to add 2).	Skill 11 (pages 64–67); Skill 12 (pages 68–70)
1.OA.6 Add and subtract within 20, demonstrating fluency for addition and subtraction within 10. Use strategies such as counting on; making ten; decomposing a number leading to a ten; using the relationship between addition and subtraction; and creating equivalent but easier or known sums.	Skill 11 (pages 64–67); Skill 12 (pages 68–70)
1.OA.8 Determine the unknown whole number in an addition or a subtraction equation relating three whole numbers.	Skill 11 (pages 64–67); Skill 12 (pages 68–70)
1.NBT.1 Count to 120, starting at any number less than 120. In this range, read and write numerals and represent a number of objects with a written numeral.	Skill 9 (pages 57–59); Skill 11 (pages 64–67); Skill 13 (pages 71–73)
1.NBT.2 Understand that the two digits of a two-digit number represent amounts of tens and ones.	Skill 9 (pages 57–59); Skill 13 (pages 71–73)
1.NBT.4 Add within 100, including adding a two-digit number and a one-digit number, and adding a two-digit number and a multiple of 10, using concrete models or drawings and strategies based on place value, properties of operations, and/or the relationship between addition and subtraction; relate the strategy to a written method and explain the reasoning used. Understand that in adding two-digit numbers, one adds tens and tens, ones and ones; and sometimes it is necessary to compose a ten.	Skill 11 (pages 64–67); Skill 12 (pages 68–70)

Correlation to Standards *(cont.)*

Grade 2	
Standard	**Skill**
2.OA.1 Use addition and subtraction within 100 to solve one- and two-step word problems involving situations of adding to, taking from, putting together, taking apart, and comparing, with unknowns in all positions (e.g., by using drawings and equations with a symbol for the unknown number to represent the problem).	Skill 16 (pages 84–86); Skill 18 (pages 91–93)
2.OA.2 Fluently add and subtract within 20 using mental strategies. By end of Grade 2, know from memory all sums of two one-digit numbers.	Skill 15 (pages 80–83); Skill 16 (pages 84–86)
2.NBT.1 Understand that the three digits of a three-digit number represent amounts of hundreds, tens, and ones (e.g., 706 equals 7 hundreds, 0 tens, and 6 ones).	Skill 19 (pages 94–96)
2.NBT.2 Count within 1,000; skip-count by 5s, 10s, and 100s.	Skill 17 (pages 87–90)
2.NBT.3 Read and write numbers to 1,000 using base-ten numerals, number names, and expanded form.	Skill 19 (pages 94–96)
2.NBT.5 Fluently add and subtract within 100, using strategies based on place value, properties of operations, and/or the relationship between addition and subtraction.	Skill 18 (pages 91–93)
2.NBT.6 Add up to four two-digit numbers, using strategies based on place value and properties of operations.	Skill 18 (pages 91–93)
2.NBT.7 Add and subtract within 1,000, using concrete models or drawings and strategies based on place value, properties of operations, and/or the relationship between addition and subtraction; relate the strategy to a written method. Understand that in adding or subtracting three-digit numbers, one adds or subtracts hundreds and hundreds, tens and tens, and ones and ones; and sometimes it is necessary to compose or decompose tens or hundreds.	Skill 18 (pages 91–93)
2.NBT.8 Mentally add 10 or 100 to a given number 100–900, and mentally subtract 10 or 100 from a given number 100–900.	Skill 18 (pages 91–93)

Correlation to Standards *(cont.)*

Grade 3	
Standard	**Skill**
3.OA.1 Interpret products of whole numbers, e.g., interpret 5 × 7 as the total number of objects in 5 groups of 7 objects each.	Skill 27 (pages 124–126)
3.OA.8 Solve two-step word problems using the four operations. Represent these problems using equations with a letter standing for the unknown quantity. Assess the reasonableness of answers using mental computation and estimation strategies, including rounding.	Skill 29 (pages 131–134)
3.NBT.1 Use place value understanding to round whole numbers to the nearest 10 or 100.	Skill 23 (pages 111–112); Skill 24 (pages 113–114)
3.NBT.2 Fluently add and subtract within 1,000, using strategies and algorithms based on place value, properties of operations, and/or the relationship between addition and subtraction.	Skill 20 (pages 99–101); Skill 22 (pages 107–110); Skill 25 (pages 115–117)
3.NF.1 Understand a fraction *1/b* as the quantity formed by 1 part when a whole is partitioned into *b* equal parts; understand a fraction *a/b* as the quantity formed by a parts of size *1/b*.	Skill 28 (pages 127–130)
3.MD.1 Tell and write time to the nearest minute and measure time intervals in minutes. Solve word problems involving addition and subtraction of time intervals in minutes (e.g., by representing the problem on a number-line diagram).	Skill 21 (pages 102–106)
3.MD.6 Measure areas by counting unit squares (square cm, square m, square in., square ft., and improvised units).	Skill 21 (pages 102–106)
3.G.2 Partition shapes into parts with equal areas. Express the area of each part as a unit fraction of the whole.	Skill 28 (pages 127–130)

About the Author

Photo Credit: Lynn Gregg

Bob Sornson, Ph.D., is nationally recognized for developing the Early Learning Success Initiative and works with schools and community organizations across the country to support teachers and parents. Dr. Sornson is dedicated to giving every child the opportunity to achieve early learning success, which lays the foundation for success in life. Bob is the author of *Stand Up and Speak Up For Yourself and Others, Stand in My Shoes: Kids Learning about Empathy, The Juice Box Bully, Fanatically Formative During the Crucial K–3 Years, Creating Classrooms Where Teachers Love to Teach, The Pre-K to Grade 3 Essential Skill Inventories,* and many other publications.

Dr. Sornson is a leader of and advocate for programs and practices that support early learning success, high-quality early childhood learning programs, and parent engagement. His workshops and keynote presentations are known for his storytelling, humor, and interactive style.

Skills and Activities for Proficiency in Pre-Kindergarten

The pre-kindergarten math skills should be learned to proficiency, a level of deep understanding that allows a student to use the skill in multiple contexts and in new situations. These four skills are not intended to be a comprehensive curriculum or an instructional program. Math learning experiences in pre-kindergarten should include numerals, measuring, balancing, movement, building, prediction, and activities and projects that bring math to life. Use rich mathematics vocabulary while engaging students in these activities (see *Math Vocabulary*, page 166). The essential skills provide a guide to a small set of skills that are indispensable, developing a basic foundation that allows students to proceed to higher levels of mathematics.

Recommended Materials

- Vinyl non-skid spots
- Beanbags
- Abacus
- Ball
- Buttons
- Hot glue
- Index cards
- Number cubes
- Counters
- Paper cups
- Manipulatives
- Letter tiles
- Blocks
- Plastic containers
- Erasers
- Puzzle pieces
- String
- Pencils
- Ribbons
- Calendar
- Stickers
- Interlocking cubes
- Pen tops
- Old keys
- Two containers
- Marbles
- Crackers, two different types
- Pictures of animals
- Dish
- Cardboard rectangle
- Tape or chalk *(optional)*

Skills and Activities for Proficiency in Pre-Kindergarten *(cont.)*

Essential Math Skills

Skill 1: Demonstrates one-to-one correspondence for numbers 1–10, with steps

Skill 2: Demonstrates one-to-one correspondence for numbers 1–10, with manipulatives

Skill 3: Adds on, using numbers 1–10, with steps

Skill 4: Adds on, using numbers 1–10, with manipulatives

Additional Resources

- *Here's What I Think* (page 148; whatthink.pdf) *(optional)*

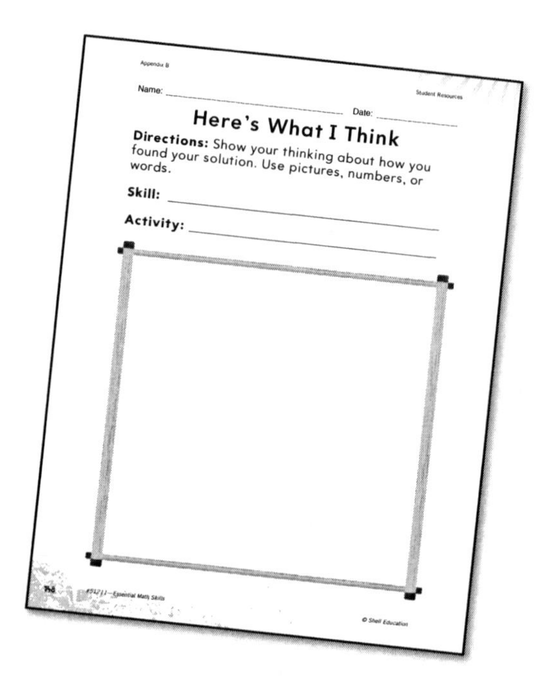
Appendix B

Student Resources

Name: ________________ Date: ________

Here's What I Think

Directions: Show your thinking about how you found your solution. Use pictures, numbers, or words.

Skill: ________________

Activity: ________________

148 #51211—Essential Math Skills © Shell Education

Pre-Kindergarten Skills

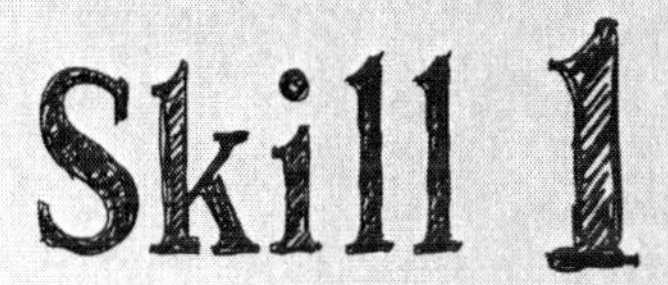

Demonstrates one-to-one correspondence for numbers 1–10, with steps

Skill Progression

Proficiency for this skill is demonstrated by the student's ability to count out loud and correctly step to the chosen number with 100 percent accuracy for each of 10 attempts on three or more days, using three or more different activities. While not all the following activities involve stepping, they do focus on gross-motor skills similar to taking steps.

Emerging	Developing	Proficient
Not yet able to count steps in sequence	Able to count 2 to 9 steps in sequence, but is inconsistent	Counts 10 or more steps in sequence

Activities for Skill Development and Assessment

Walk the Line

Place 10 vinyl non-skid spots in a line on the floor. Use a marker or stickers to add 1–10 dots to each spot. Place the spots on the ground, in order. Ask the student to stand on the starting spot. Choose a number between 1 and 10, and model taking that number of forward steps on the spots. Then, ask the student to take any number (1–10) of forward steps, stepping on the next group of dots with each step. For example, say, "Take four steps forward." As the student steps, have him or her count out loud. Practice 5–10 times per day. Give positive feedback each time the student performs the activity. For example, "Great job. You stopped on the four dots! Now, go back to the start. Try six steps forward." Allow the student to notice the pattern of dots on the floor without much explanation from you. Celebrate successes.

Teacher Tip: A number line can be made on the floor with tape, drawn with chalk on concrete or asphalt, or painted on non-skid carpet squares.

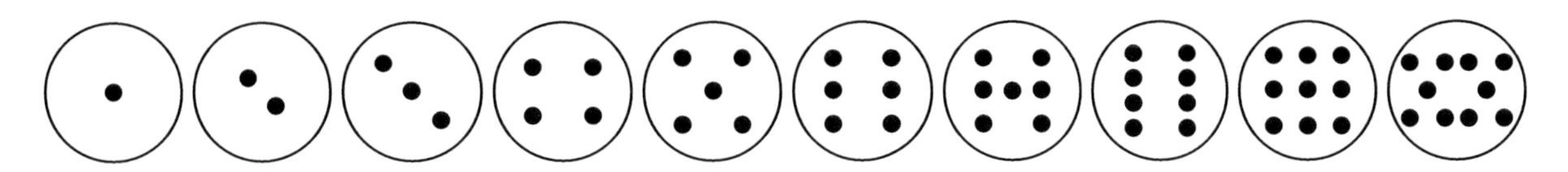

Stepping Stones

Place 10 vinyl non-skid spots on the floor as "stepping stones" (not in a straight line). Say, "Take _______ steps forward." Practice 5–10 repetitions of this activity in a session, with great enthusiasm for each performance. For example, say, "Great job. You are standing on the five dots. Now, go back to start. Ready? Now count and step to eight." Vary the difficulty by your placement of the stepping stones. Stretching, crossing midline, and varying distances between "stones" will help the student internalize the sequence and value of numbers. Students also have to actively look for the next stepping stone to determine where to step.

Go the Distance

Identify a destination (e.g., a blue chair, a door) and instruct the student to count aloud his or her steps from a starting point to the destination. Begin with easy missions. For example, "Count aloud your steps from here to the door." Use short distances of 10 steps or fewer. Repeat the activity 5–10 times in a session so that the student can feel the difference between varying distances.

Variation: Ask the student to count sideward steps or backward steps to reach a destination. Changing the direction of the steps makes the activity seem like a new challenge and keeps students engaged.

Ups and Downs

Practice counting steps on a stairway. Standing at the bottom of the stairs, throw a beanbag partway up the flight of stairs. Ask the student to count the steps as he or she goes to retrieve the beanbag. Repeat the activity 5–10 times in a session so that the student can feel the difference between varying numbers of steps.

Activities to Extend Deep Understanding

Catch

Ask the student to stand and catch a certain number of beanbags. Make this activity fun and easy by throwing the beanbags gently, underhand, directly toward the student's chest. Encourage the student to catch with two hands and then drop the beanbag. When he or she has caught the correct number of beanbags, the student should say *stop*. Some students will count aloud, while others will count silently before saying *stop*. Repeat this activity 5–10 times in a session.

Variation: Catch and Hold—Have the student catch and hold the beanbags until he or she is holding the correct number.

Throw

Ask the student to throw a certain number of beanbags into a crate or toward a target. Have the student count only the beanbags that land in the target. Celebrate successes! Repeat this activity 5–10 times in a session.

Imitation

At the beginning of a whole-group activity, ask students to repeat a movement that you make a certain number of times. You might nod your head three times and ask the students to nod three times.

Follow Me

Swing your arms as you step and count out loud, having students follow and count with you. Stop walking at various numbers. Pause, and then start the process over again, beginning with *one*.

Estimation

Show students how to carefully toss a beanbag underhand, guess the distance to it in steps, and then count the steps it takes to get to the beanbag. Give a student a beanbag. Ask him or her to toss it, guess the distance to it in steps, and then count the actual number of steps.

On Tour

Take a tour around the school. Stop at various places, and ask students to guess how many steps it will take to reach a specific destination such as a stairway, a fire extinguisher, an office door, or a display case. Then, test the students' estimates.

Setting Up

When putting out carpet squares or chairs, ask students to stand beside them and count students and the seats, emphasizing one seat for one student.

Pre-Kindergarten Skills

Skill 2

Demonstrates one-to-one correspondence for numbers 1–10, with manipulatives

Skill Progression

Proficiency for this skill is demonstrated by the student's ability to count objects out loud correctly to the chosen number with 100 percent accuracy for each of 10 attempts on three or more days, using three or more different activities. The activities provided for this skill focus on fine-motor skills.

Emerging	Developing	Proficient
Not yet counting objects with one-to-one correspondence	Counts objects with accuracy to 3	Shows one-to-one correspondence when counting 10 or more objects

Activities for Skill Development and Assessment

Touch and Count

Put several small objects (e.g., marbles, buttons, small stones, pencils, clothespins) into a dish. Ask the student to count aloud while moving the objects from one dish to another. This exercise supports the relationship between numerals and quantity and provides visual, auditory, and kinesthetic experiences. Repeat this activity 5–10 times in a daily session.

Variation: Have the student estimate before counting, and then count and move the objects.

Absolutely Abacus: How Many?

Place the abacus in front of the student. Using the top row of beads, move a certain number of beads to one side. Ask, "How many is this?" Allow the student to count the beads aloud, one by one. Use numbers that allow the student to quickly find success at least 95 percent of the time. Repeat this activity 5–10 times in a daily session.

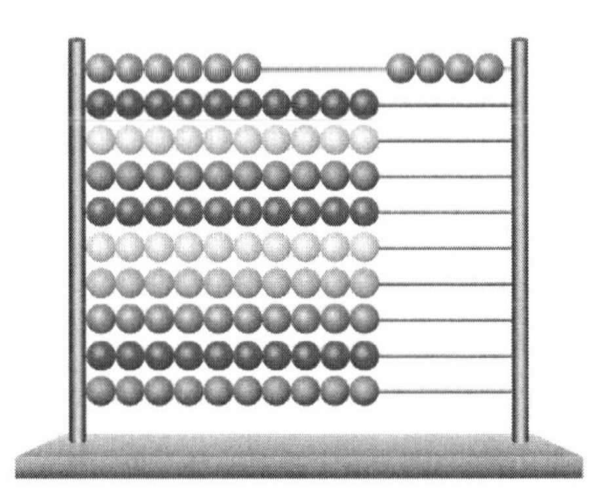

Toss Up

Ask the student to toss a ball or a beanbag up in the air up to 10 times, counting with each catch. Keep it fun and successful.

Button Cards

Use hot glue to attach buttons to index cards. Create cards with varying numbers of buttons, up to 10. Flash a card and ask the student to estimate how many buttons are on the card. After the student has estimated, allow him or her to count the buttons to check the estimate.

Fill the Cup

Give the student a number cube and 10 counters. Tell him or her to roll the number cube and count the number rolled. Then, have the student count out that number of counters and put them into a paper cup.

Variation: Use different manipulatives (e.g., interlocking cubes, pasta, beans) to keep the activity interesting for the student.

Counting Letters

Using letter tiles, help the student to spell his or her first name. Ask the student to count the letters. Continue spelling out other names, such as those of siblings, parents, classmates, or pets, and ask the student to count the letters in each name.

Junk in the Trunk

Collect items from around the classroom (e.g., erasers, crayons, blocks, puzzle pieces) and put them into a plastic container. Plan to collect one button, two blocks, three erasers, four puzzle pieces, five paper clips, six bits of string, seven pencils, eight ribbons, nine pen tops, 10 old keys, or any similar combination. Ask the student to sort the items into piles of the same items, and then count how many are in each pile.

Tapping Out

Model counting while tapping on your leg up to 10 times. Then, ask the student to copy your pattern. Sitting across from the student, model a simple one-leg counting pattern (e.g., tapping with one hand on one leg while counting out loud). Vary rhythm and complexity as long as the student is having fun and is successful at least 95 percent of the time. Introduce two-handed patterns (up to 10) when the student is ready.

Activities to Extend Deep Understanding

Block Measurement

Give the student a supply of identical objects (e.g., chenille stems, unused erasers, drinking straws, blocks). Ask him or her to guess and then measure how many objects it will take to cover the distance from one point to another (e.g., from one end of a desk to the other, across the cover of a big book, or from one side of a doorjamb to the other). Model how to accurately measure by properly lining up the objects. Have the student count each object he or she has used to measure.

Fill It Up

Ask the student how many beanbags (or blocks, erasers, pencils, action figures, etc.) he or she thinks will fit into a small box or bag, and then ask him or her to fill the container to capacity and count the total number of objects.

Snack-Time Counting

Ask the student to count the number of students at a table. Then, have him or her count out the number of snacks needed so that each student gets one.

Calendar Counting

Use a calendar to count days. Point to and count all the days in one week, the school days left in the week, or the number of Saturdays in the month.

Number and Value

Arrange a row of 3–5 stickers. Arrange a second row with the same number of stickers but with more space between each sticker. Ask the student whether the number of stickers in each row is the same or whether there are more or fewer. Ask the student how he or she can find out, and then allow the student to figure out how to count and compare the two rows of stickers.

Matched Sets

Create a group of 2–10 familiar objects, such as stuffed bears, baseballs, or beanbags. Using a different familiar object, ask the student to create a new set with the same number of objects as the first set.

Value to Value

Build a connected set of 3–5 interlocking cubes. Ask the student to guess how many individual cubes it will take to build another set of the same size. After the student's guess, allow him or her to test the estimate by building a new set.

Pre-Kindergarten Skills

Skill 3

Adds on, using numbers 1–10, with steps

Skill Progression

Proficiency for this skill is demonstrated by the student's ability to take the correct number of steps and then add on steps correctly and identify the final number value with 100 percent accuracy for each of 10 attempts, on three or more days, using two or more different activities. The activities provided allow the student to learn the value (through distance) of a number and learn to compare the value of numbers physically and mentally.

Emerging	Developing	Proficient
Unable to add on numbers using steps on a number line without recounting (1–10)	Inconsistently adds on numbers using steps on a number line without recounting (1–10)	Adds on numbers using steps on a number line without recounting (1–10)

Activities for Skill Development and Assessment

Walk On

Use vinyl spots, tape, or chalk to create a number line on the floor. Ask the student to move to a certain number, stepping on each number along the way. Once there, ask the student to take a certain number of additional steps and then ask which number he or she is standing on. For example, "Step to five. Great job! Now, take three more steps. Which number are you standing on?" The student should give you the number value of the spot on which he or she is standing.

Variation: Use numerals rather than dots on the number line. This helps the student identify numerals and use them to represent a number value. The student will tell you when this activity gets too easy.

Estimation for Education

Ask the student to walk to a destination within the room, counting the steps to get there. Choose a destination 6–10 steps away. Then, direct the student to walk partway back and stop. Don't ask the student to count steps during this phase of the activity; see whether he or she does it independently. Now, ask the student to estimate how many steps it will take to get to the original location. After the student estimates, ask him or her to walk back and check the estimation. Show enthusiasm when the student makes a good estimate.

Continue the Climb

Using a stairway, ask the student to count the steps from bottom to top. Ask the student to take a certain number of steps up the stairs. Then, ask him or her to proceed a certain number of additional steps and identify where he or she stands.

Teacher Tip: Label the steps with vinyl spots marked with dots (see page 23) to make the activity easier initially.

Horizontal Ladder

If the school playground has a play structure with a horizontal ladder, ask the student to count the bars from one end to the other. Ask the student to swing to a certain bar, for example, the fourth bar. Then, ask him or her to swing forward a certain number of bars and identify the value of the bar from which he or she is hanging. Arm strength is required for this activity; carefully spot the student.

Everglades Trek

Ask the student to pretend the playground is the Everglades, teeming with alligators. Place vinyl spots marked with dots (see page 23) on the ground as "safety spots," just far enough apart so that the student must stretch to reach each one. Ask the student to move to a certain number, stepping on each number along the way. Once there, ask the student to take additional steps forward, and then ask, "Where are you now?" The student will give you the number value of the spot on which he or she is standing.

Walk Back *(Challenge Activity)*

Using a number line (created with dots, not numerals), direct the student to take a certain number of steps forward. Then, have him or her take a certain number of steps backward. Ask, "Where are you now?"

While this activity is easy for many students, continue to practice 5 repetitions per day for at least 20 days. Although you are teaching the concept of subtraction, never describe that concept to the student. Allow the student to enjoy the confidence with which he or she can be successful and to develop a kinesthetic awareness of moving from a larger number to a smaller number by stepping backwards.

Variation: After the student seems to understand all the number values, use numerals rather than dots on the number line. This helps the student to symbolically represent a number value.

Pre-Kindergarten Skills

Skill 4

Adds on, using numbers 1–10, with manipulatives

Skill Progression

Proficiency for this skill is demonstrated by the student's ability to count the correct number of objects, then add on correctly and identify the final number value with 100 percent accuracy for each of 10 attempts on three or more days, using three or more different activities.

Emerging	Developing	Proficient
Unable to add on numbers without recounting (1–10)	Inconsistently adds on from a set of objects without recounting (1–10)	Adds on from a set of objects without recounting (1–10)

Activities for Skill Development and Assessment

Absolutely Abacus: Show Me

Give the student an abacus or a counting frame and have him or her show you a certain number of beads on the first row. Allow plenty of time for the student to move the beads, as many young students may have trouble at first. Then, tell him or her to add a certain number more to that group. Ask the student how many beads there are now altogether. Play with adding on within the row of 10 beads. For example, you might say, "Show me five beads. Good job. Now, add on two more. How many are there now?"

Summation

Show the student two containers, each filled with a single color of marbles (e.g., one container of blue marbles and one container of yellow marbles). Ask the student to pick a certain number of marbles from each container. Then, ask him or her to determine how many total marbles he or she has. Give the students problems that can be solved within about 10 seconds.

Variation: Replace the marbles with plastic chips, interlocking cubes, or other small objects in order to keep the activity engaging.

Snack Attack

At snack time, lay out a certain number of one type of cracker (e.g., square crackers), and ask the student to count the crackers. Then, display a certain number of a different type of cracker (e.g., round crackers), and have the student add on. Ask how many crackers there are altogether.

Interlocking Cubes

Ask the student to put together a certain number of interlocking cubes. Then, ask him or her to add a certain number more to that group. Ask the student how many cubes there are now altogether. Play with adding on within a total of 10. For example, "Put together four cubes. Good job. Now, add on two more. How many are there now?"

Button Up

Show the student a button card with up to 5 buttons on it. Ask the student to count the buttons. Then, show another button card and encourage the student to add on. Ask the student to determine the total number of buttons.

Loads of Legs

Gather a collection of pictures of creatures that have different numbers of legs (e.g., people, dogs, spiders, insects). Have the student choose one picture and tell you how many legs the creature has (he or she can count the legs if necessary). Then, have the student choose another picture and again tell you how many legs he or she sees. Finally, ask the student to tell you how many legs there are altogether in the two pictures. Continue the activity with two new pictures.

Junk in a Jar

Give a student a clean plastic jar and 10 marbles (or use blocks, buttons, beads, bells, etc.). Tell the student to place a specific number of marbles (1–10) into his or her jar. Next, ask the student how many marbles would be in the jar if a certain number more were added. After the student adds these marbles, check the sum.

Conservation of Number Value *(Challenge Activity)*

Place a set of objects (e.g., buttons, beads, marbles, blocks, crayons, pencils) in your hands. Have the student count the number of objects in your hands. Add 1–3 more objects to the set in your hands. Allow the student to observe and understand how many were added to the group without letting him or her see the new group. Then, ask the student to tell you how many are now in your hands.

Place a set of objects on a dish. Cover the dish with a piece of cardboard to screen the student's view of the objects. Add 1–3 objects, allowing the student to clearly see how many were added or taken away but not letting him or her see the new set. Again, ask the student to tell you how many are now in the group. Try the same activity using objects in a hat, in a shoe box, in a cupboard, or under a blanket.

Teacher Tip: Do not exceed 10 objects in a set for this activity. Use number values that allow each student to experience great success with this activity. Be cautious not to make it too difficult.

Skills and Activities

for Proficiency in

Kindergarten

In kindergarten, the goal is to help students deeply understand the basic number values and concepts that make math fun. In some schools, the tendency to race quickly through too much content causes students to become frustrated, fail to deeply understand essential concepts, and begin to memorize answers that they do not fully understand.

The kindergarten math skills should be learned to proficiency—a level of deep understanding that allows a student to use the skills in multiple contexts and in new situations. These four skills are not intended to be a comprehensive curriculum or an instructional program. Math learning experiences in kindergarten should include numerals, measuring, movement, prediction, basic addition and subtraction, and activities and projects that bring math to life. Use rich mathematics vocabulary on word walls and while engaging students in these activities (see *Math Vocabulary,* page 166).

The essential skills provide a guide to a small set of skills that are indispensable—a basic foundation that allows students to proceed to higher levels of mathematics.

Recommended Materials

- Calendar
- Counters
- Vinyl non-skid spots
- Number tiles
- Tagboard
- Beads
- Buttons
- Erasers
- Stuffed animals
- Pennies
- Number cubes
- Pattern blocks
- Rubber stamps and ink pad
- Construction paper (4 colors)
- Stackable blocks (4 colors)
- Cups (4 colors)
- Clothespins (4 colors)
- Clothesline
- Keyboard
- Abacus
- Small stones
- Letter tiles
- Paper bags
- Empty food containers
- Beanbags
- Crate, hoop, or tape
- Crackers
- Interlocking cubes
- Marbles
- Bowl
- Jars
- Dominoes
- Socks (different colors)
- Craft sticks
- Chenille stems
- Large beads
- Playing cards

Skills and Activities for Proficiency in Kindergarten *(cont.)*

Essential Math Skills

Skill 5: Demonstrates counting to 100

Skill 6: Has one-to-one correspondence for numbers 1–30

Skill 7: Understands combinations (within 10)

Skill 8: Recognizes number groups without counting (2–10)

Additional Resources

- *Ten Chart* (tenchart.pdf)
- *Twenty Chart* (twentychart.pdf)
- *Fifty Chart* (page 152; fiftychart.pdf)
- *Hundred Chart* (page 150; hundredchart.pdf)
- *Here's What I Think* (page 148; whatthink.pdf) *(optional)*

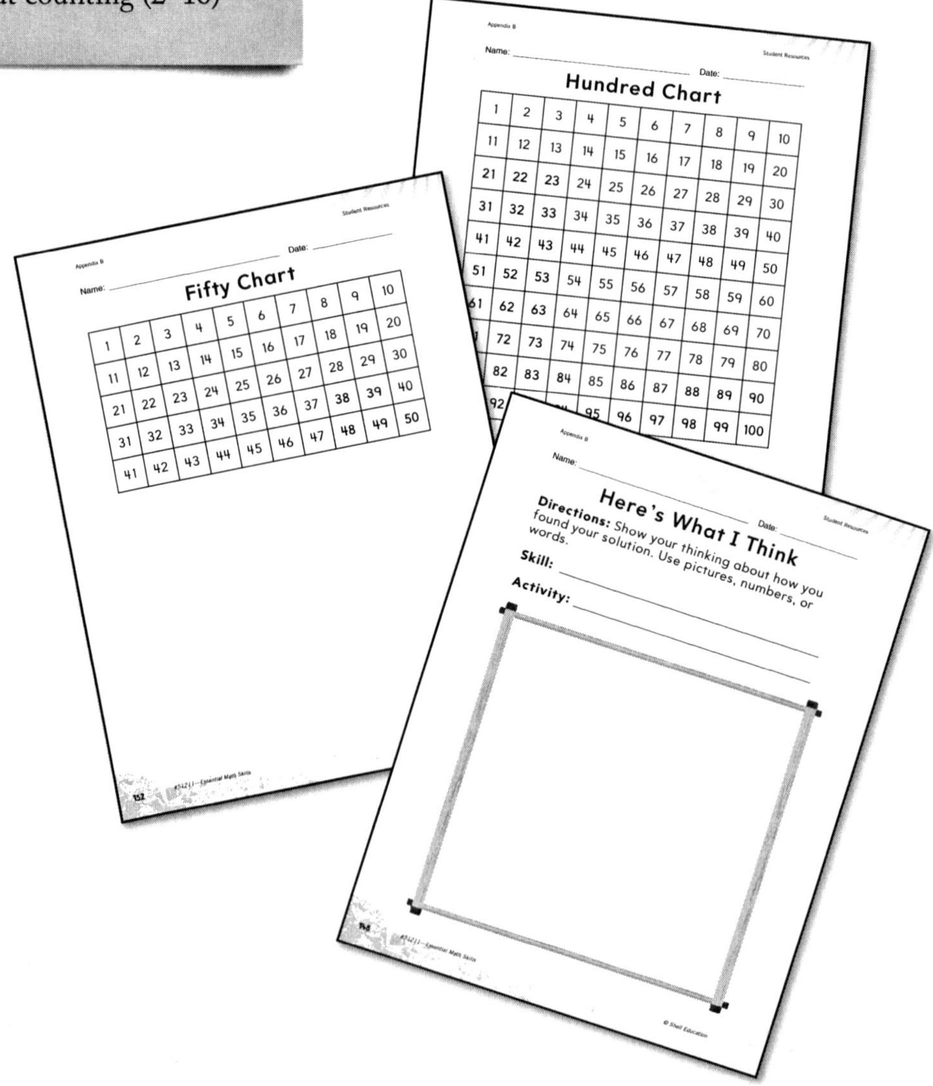

Kindergarten Skills

Skill 5

Demonstrates counting to 100

Skill Progression

Students typically learn to say the correct sequence of numbers before fully understanding the value of each number. This is a normal part of learning to enjoy the use of numbers. Learning to count is one of the essential early math skills, and is part of making math fun when combined with activities that help students understand the value of numbers. Proficiency for this skill is demonstrated by the student's ability to easily count in sequence to 100.

Intervention	Developing	Proficient
Counts to a number less than 30 with accuracy	Counts to 31–99 with accuracy	Counts to 100 with accuracy

Activities for Skill Development and Assessment

Marching Numbers

Lead the student in counting to 10 or higher (up to 100) while marching in place or around the room. Help the student find the connection between counting and marching.

Variation: Stepping Out—Count out loud as the class walks to various locations around the building, allowing students to hear the cadence and sequence of counting.

Dancing Hands

Count aloud to 10 or higher while beating the table with open hands. Try counting to different numbers with different speeds and beat patterns.

Variation: Dancing Fingers—Count to 10 or higher while touching the table with each finger in sequence.

Calendar Counting

Have the student practice counting the number of school days in the month and the number of total days in the month while referring to a large classroom calendar.

Number Chart Counting

Using a *Ten Chart* (tenchart.pdf), a *Twenty Chart* (twentychart.pdf), a *Fifty Chart* (page 152), and eventually a *Hundred Chart* (page 150), count aloud with the student to various number values.

Chart It Out

Using a *Ten Chart* (tenchart.pdf), a *Twenty Chart* (twentychart.pdf), a *Fifty Chart* (page 152), or a *Hundred Chart* (page 150) and counters, give the student a number value, show him or her where it is on the chart, and have the student fill in the correct number of counters on the number chart.

Connect the Dots

Place stepping spots with dots in various locations in the room. Ask the student to find them in the correct order.

Number-Value Puzzles

On a strip of tagboard, draw groups of dots representing number values 1–10. Cut the numbers apart into puzzle pieces. Allow the student to solve the puzzle, putting the number values in the correct sequence.

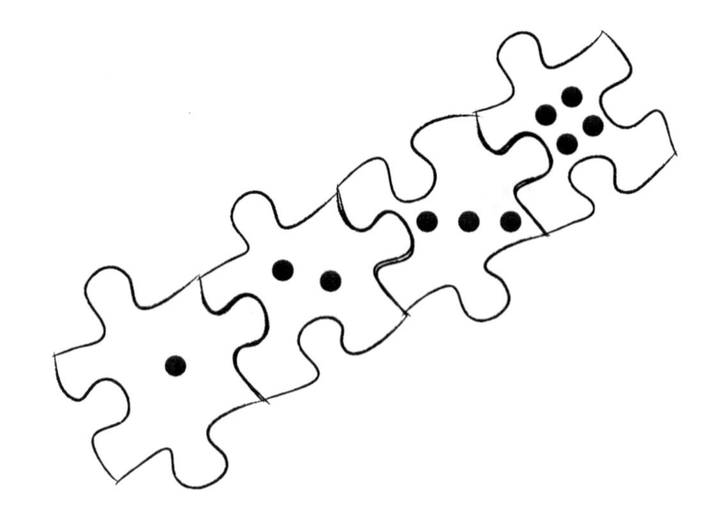

Number-Order Puzzles

On a strip of tagboard, write the numerals 1–10, 1–20, or 1–30. Then, cut the strip into puzzle pieces. Allow the student to solve the puzzle and put the numbers in the correct sequence.

Scrambled Numbers

Provide the student with a set of number tiles with values (represented as dots or numerals) up to a value to which he or she already knows how to count. Scramble the tiles and have the student put them in sequential order, and then count them aloud.

Count On

Choose a number within the student's ability to accurately count, and then ask him or her to count on from that number to another number.

Activities to Extend Deep Understanding

More and Less

For many students, the concepts of *more* and *less* may be unfamiliar. Allow plenty of time for the student to internalize the meanings of *more* and *less*.

- Using real objects (e.g., stuffed animals, blocks, paper clips, buttons, beads, or pennies), show two groups of objects to the student and have him or her identify which is more or less.
- Using picture cards of groups of objects, show two cards to the student and have him or her identify which is more or less.
- Clap your hands a certain number of times and then pause and clap a different number of times. Have the student identify which set had the greater number of claps.
- Tap on a table a certain number of times and then pause and tap a different number of times. Have the student identify which set had the greater number of taps.
- Ask the student to roll two number cubes and then identify which value is greater. Have the student try rolling three number cubes and then put them in order from greatest to least.

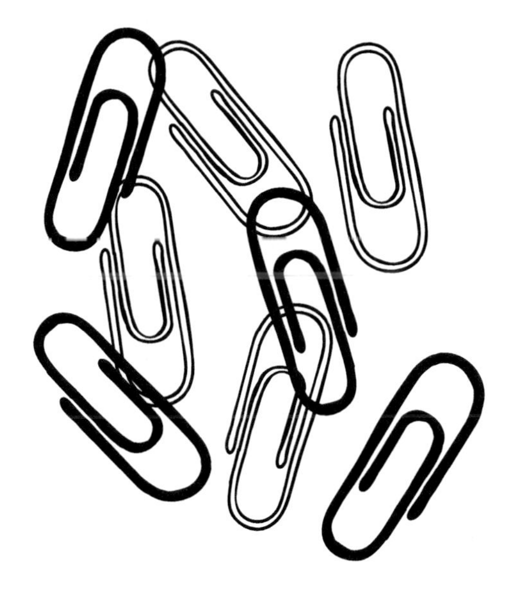

Odd and Even

- Using real objects (e.g., stuffed animals, blocks, erasers, buttons, beads, or pennies), teach the student to understand the concepts of matching pairs and odd or even numbers. Then, play with groups of objects to practice identifying which sets of objects represent odd or even number values.
- Using picture cards of familiar objects, have the student identify which sets of objects represent odd or even number values.
- Call out a number 1–10 (for example, 4). Teach the student how to use alternating hands to count. Have the student start counting on his or her right thumb, saying *one*; on his left thumb, saying *two*; on his right index finger, saying *three*; on his left index finger, saying *four*. When the student has reached the designated number, the student should put his or her hands together to see whether every finger has a partner. If it does, the number is even; if it doesn't, the number is odd.

- Select a random group of number tiles within the range of the student's ability to easily count. Ask him or her to identify and group the odd and the even numbers.
- Clap your hands a certain number of times. Have the student identify whether the number of claps was odd or even.
- Put one hand behind your back and display any number of fingers. Have the student guess whether you are showing an odd or an even number. Then, show your hand so that the student can check his or her guess.

Patterns

Try modeling two-part tasks such as patting your head and then touching both knees. Once the student is successful with two-part tasks, try three-part gross-motor patterns such as stepping forward and then tapping both knees with your hands, and then clapping hands over your head. Again, ask the student to copy you. As the student becomes skilled at copying three-part patterns, challenge him or her by adding a fourth task.

- Create a two-part, a three-part, or a four-part pattern, using pattern blocks. Have the student extend the pattern until it has been repeated three times.
- Try creating patterns by tracing pattern blocks onto strips of paper. Then, have the student extend the pattern until it has been repeated three times.
- Using rubber stamps, ink pads, and some strips of paper, begin a three-part or four-part pattern using the stamps. Ask the student to continue the pattern you started.
- Using construction paper in at least four different colors, build a paper chain in a three-part or four-part pattern. Have the student extend the pattern.
- Using a collection of large colored blocks, build a tower of stacked blocks in a three-part or four-part color pattern. After you have built it so the pattern begins to repeat, have the student take over and continue the pattern as high as possible.
- Gather stacks of cups in at least four different colors. Create a pattern of stacked cups, and have the student replicate the pattern.
- Gather an assortment of at least four different colored plastic clothespins (or use permanent markers to color the ends of wooden clothespins). On a clothesline, create a pattern on the line, and have the student extend the pattern.
- Using a keyboard, model a three- or four-step pattern and ask the student to repeat the pattern. Experiment with patterns and complexity.

Skill 6

Has one-to-one correspondence for numbers 1–30

Skill Progression

Proficiency for this skill is demonstrated by the student's ability to count any group of objects or steps with accuracy to 30 on three or more days, using three or more different activities.

Intervention	Developing	Proficient
Can count fewer than 10 objects with one-to-one correspondence	Shows one-to-one correspondence when counting 10–29 objects	Shows one-to-one correspondence when counting 30 or more objects

Activities for Skill Development and Assessment

Go the Distance

Identify a destination that is more than 10 steps away. Then, have the student count aloud as he or she steps to that destination. Practice 5–10 repetitions in a session.

Variation: Count steps on a stairway. The extra effort of climbing stairs adds one more element to kinesthetic understanding. Another variation is using sideways or backward steps.

Stepping-Stone Pathway

Create a stepping-stone pathway to 30 in the classroom or outside. Use vinyl spots marked with dots or numerals as the stepping stones. You might purposely replace some of the spots with blank stepping stones so that the student does some thinking about the value of that spot. Have the student practice counting aloud while walking forward, walking backward, or counting on.

Absolutely Abacus: Show Me

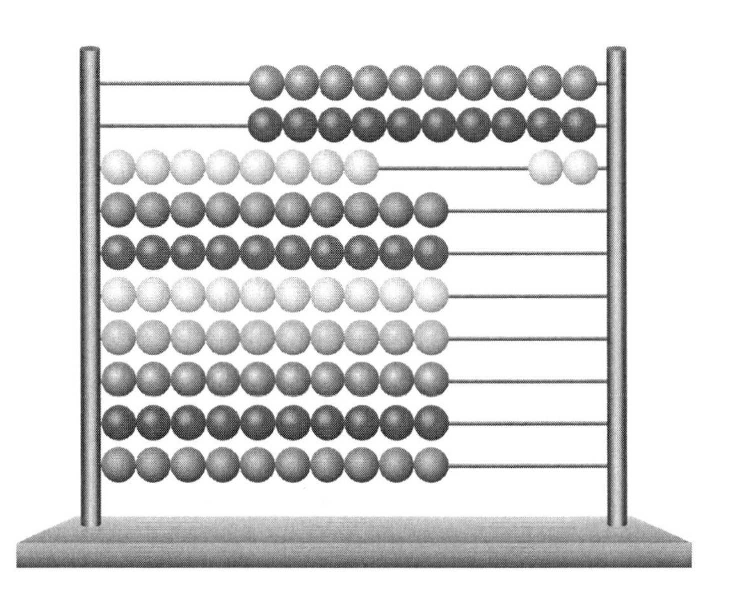

Place the abacus in front of the student. Ask the student to show you a certain number (up to 30) on the abacus. Allow plenty of time for him or her to move the beads, as many young students may have trouble at first. Encourage the student to correctly count out the number, using the top row of the abacus before moving on to the next row. Allow the student to discover patterns of numbers without too much coaching from you. Repeat this activity 10–20 times in a session.

Touch and Count

Put 5–30 objects (e.g., marbles, buttons, small stones, pencils, clothespins) into a dish. Ask the student to count while transferring the objects to another dish. This exercise supports the relationship between spoken numbers and quantity. It is visual, auditory, and kinesthetic. Have fun! Repeat this activity 5–10 times in a session.

Variation: Have the student estimate before counting, and then count and move the objects.

Number Cubes It Up

Give the student two number cubes. Have him or her roll the number cubes and count the sum. As his or her counting skills develop, add a third or a fourth number cube.

Variation: Combine this activity with the abacus so the student can see the number values as he or she finds the sum.

Counting Letters

Using letter tiles, spell out a short sentence containing familiar words. Have the student count the number of letters in the sentence.

Estimation

Collect up to 30 items from around the classroom (e.g., erasers, crayons, blocks, puzzle pieces) and put them into a plastic container. Ask the student to first estimate how many objects are in the container and then count the total. Then, have the student sort the items into piles of similar items and then count the number of items in each group. Change the total number of items in the container for future sessions.

Absolutely Abacus: Counting On

Place the abacus in front of the student. Show the student a certain number on the abacus. Then, ask him or her to add beads one at a time, and count on from this number up to 30.

Unloading Groceries

Using paper grocery bags and a collection of artificial food items or small, empty food containers (e.g., soup cans, yogurt cups, small cereal boxes, juice boxes), ask the student to count the items in a bag. Begin with just a few items and then increase the number as the student becomes more skilled.

Kindergarten Skills

Skill 7

Understands combinations (within 10)

Skill Progression

Proficiency for this skill is demonstrated by the student's ability to add on to or take away from a set of objects without recounting, using two numbers with a combined value up to 10, on three or more days, using three or more different activities.

Intervention	Developing	Proficient
Does not consistently understand how to add on 1–3 objects without recounting	Consistently adds on 1–3 objects without recounting	Adds on to or takes away from a set of objects without recounting (1–10)

Activities for Skill Development and Assessment

Walk On Combinations

Use vinyl spots to create a number line on the floor (with dots in a pattern as found on dice or dominoes, but no numerals). Ask the student to move to a certain number, stepping on each number along the way. Once there, ask the student, "How many more steps will it take to get to [a number]?" Allow the to student observe, think, and then respond. Encourage the student to check his or her answer. Repeat this activity five or more times in a session.

Variation 1: Ask the student to move to a certain number, stepping on each number along the way. Once there, say, "Imagine you take [a number] steps *more*. Where will you be?"

Variation 2: Ask the student to move to [a number], stepping on each number along the way. Once there, ask the student, "Imagine you take [a number] steps *back*. Where will you be?"

Variation 3: Use numerals rather than dots for the number line. This helps the student identify numerals and use them to represent a number value.

Estimation for Education

This activity develops an awareness of number combinations through estimation and movement. Ask the student to walk to a destination within the room, counting the steps. Choose a destination 6–10 steps away. Then, direct him or her to walk partway back and stop. Don't tell the student to count the steps during this phase of the activity. See if he or she does it independently. Now, ask the student to estimate how many more steps it will take to get to the original location. After the student estimates, ask him or her to walk back and check the estimation. Get excited when the student estimates correctly!

Target Toss

Establish a target that can be reached somewhat easily with a gentle toss. The student will occasionally miss. The target could be a crate, a hoop, or a shape taped to the floor. Mark the tossing location on the floor. Have 10 beanbags available. Ask the student to pick up a certain number of beanbags and then toss them underhand, one at a time, toward the target. Avoid making the target too difficult to hit. Ideally, 75–90 percent of the student's tosses will be on target. Success keeps the activity fun for the student. Once all beanbags are tossed, ask the student to identify how many are on the target and how many are not. Don't require the student to add the two numbers to get a total; let him or her make the connection independently.

Cracker Combinations

Lay out two kinds of crackers and display different combinations of the two together. Ask the student to tell you how many there are of each kind and how many there are altogether.

Interlocking Cubes

Ask the student to put together a certain number of interlocking cubes. Then, tell him or her to add a certain number more to that set. Ask the student how many cubes there are altogether. Play with adding on and taking away, using combinations that don't exceed 10. For example, you might say, "Put together four cubes. Good job! Now, add on two more. How many are there now?" or "Show me six cubes. Take away two cubes. How many are there now?"

Button Up

Display two button cards (see page 28), and then ask the student to determine the total number of buttons. On a small whiteboard, draw out the correct answer, using small circles to represent the number. Set the cards and whiteboard in a row to model a written equation, showing that the first card and the second card add up to the total on the whiteboard.

Variation: Ask the student to look at two button cards and then draw out the correct sum of circles on the whiteboard.

Calendar Sums

When working with the calendar, ask the student how many days he or she has come to school this week and how many school days are left in the week. Ask the student to determine how many school days there are altogether.

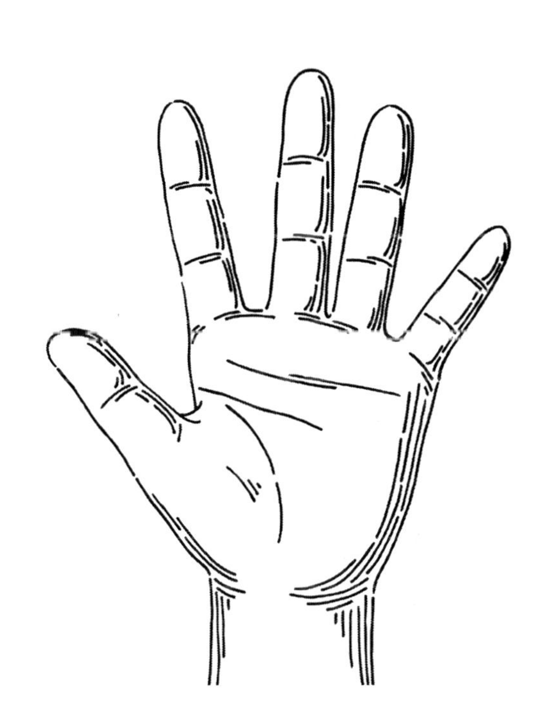

Flash Fingers

Flash a certain number of fingers on one hand and a certain number of fingers on the other hand. Ask the student to quickly flash the total number of fingers displayed. The student can use his or her fingers in any combination that correctly describes the total.

In the Hand Combinations

Hide several marbles (or other small objects) in each hand so that the sum of the marbles is no more than 10. Show the student the marbles in one hand, and then close your hand. Show the marbles in your other hand, and then close that hand. Ask the student to name the sum and then count out that number of marbles into a bowl.

Variation: Show the student the marbles in one hand and then the other hand. Ask him or her to name the sum and count that number of marbles into a bowl. Keep your hands closed. Then, show the student the marbles in one hand and ask him or her to determine how many marbles are in the other (closed) hand.

Marbles in a Jar

Give the student a jar and 10 marbles (or other small objects). Tell the student to place a specific number of marbles (1–10) in the jar. Next, ask him or her how many marbles would be in the jar if a certain number of additional marbles were added. After the student has added these marbles, have him or her check the sum.

Conservation of Number Value *(Challenge Activity)*

Place a set of objects (e.g., buttons, beads, marbles, blocks, crayons, pencils) in your hands. Have the student count the number of objects in your hands. Add a few (1–3) more objects into the set in your hands. Allow the student to observe and understand how many were added to the group, without letting him or her see the new group. Then, ask him or her to tell you how many are now in your hands.

Variation 1: Place a set of objects (e.g., buttons, beads, marbles, blocks, crayons, pencils) in your hands. Have the student count the number of objects in your hands. Take away a few (1–3) objects from the set in your hands. Allow the student to observe and understand how many were taken from the group, without letting him or her see the new group. Ask the student to tell you how many are now in your hands.

Variation 2: Place a set of objects into a dish. Cover the dish with a piece of cardboard used to screen the student's view of the objects. Add or take away 1–3 objects, allowing the student to clearly see how many were added or taken away but not letting him or her see the new set. Again, ask him or her to tell you how many are now in the group.

Variation 3: Try the same activity using objects in a hat, in a shoe box, in a cupboard, under a towel, etc. Do not exceed 10 objects in a set for this activity. Use number values that allow the student to experience great success with this activity. Be cautious not to make it too difficult.

Recognizes number groups without counting (2–10)

Skill Progression

Proficiency for this skill is demonstrated by the student's ability to quickly name any set of objects representing a number value from 2 to 10, on three or more days, using three or more different activities.

Intervention	Developing	Proficient
Recognizes number groups of 1 or 2 without counting individual objects	Recognizes number groups of 3–5 without counting individual objects	Recognizes number groups up to 10 without counting individual objects

Activities for Skill Development and Assessment

Absolutely Abacus: How Many?

Place the abacus in front of the student. Using the top row of 10 beads, move a certain number of beads to one side. Ask the student, "How many is this?" At first, he or she will count the beads one by one. With practice, the student will learn to recognize groups of objects without counting the individual objects. Do not try to rush this development. Make your sessions fast and fun. Let the student discover the patterns of numbers. Repeat this activity 10–20 times in a session.

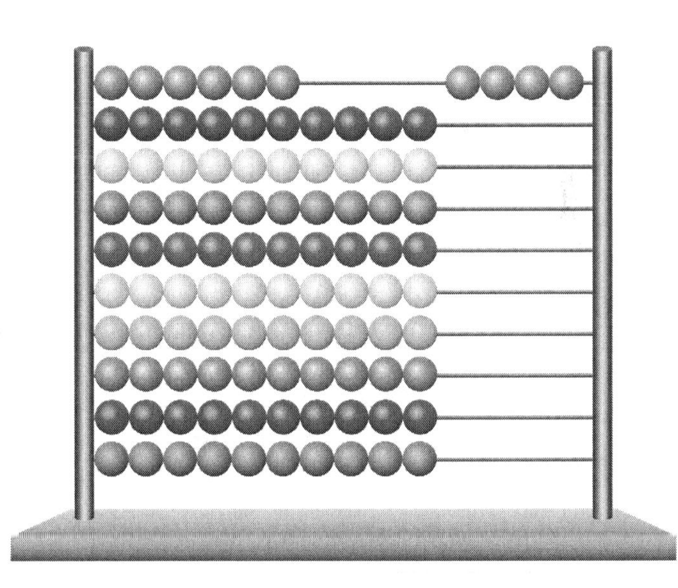

Button Flash

Flash a button card (see page 28) and ask the student to quickly tell you how many buttons were on the card. Keep this activity fast paced and fun!

Dominoes and Number Cubes

Give the student a number cube. Ask him or her to roll the number cube and quickly name the number value. Try a similar activity by drawing a domino from a bowl and naming the value on one half.

Counting Letters

Use letter tiles to spell the student's first name. Ask him or her to quickly determine the number of letters. Continue spelling other words, and ask him or her to decide as quickly as possible how many letters are in each word.

In-the-Hand Number Values

Hold a small group of marbles (no more than 10) in your hand, covering them with your fingers. Slowly open your fingers to reveal the marbles and then quickly cover them again. Ask the student to identify the number of marbles.

Absolutely Abacus: Subitizing

Place the abacus in front of the student. Using the top row of 10 beads, ask the student to move a certain number to the opposite side. Start with smaller numbers so that the student is successful right away. At first, he or she may move the beads one by one. With practice, the student will start to move them in segments or as a whole group. Resist the temptation to show him or her the patterns (e.g., 9 is always one less than 10). Allow the student to discover these patterns. Repeat this activity at least 10 times in a session.

Sort and Name

Collect a tub of different-color socks (or blocks, erasers, pencils, crayons, etc.). First, have the student sort the items by color and then quickly determine the number of items in each group. Be sure to have no more than 10 of each color.

Grab and Go

Lay out 10 craft sticks in front of the student. Say a number from 1–10, and then ask the student to quickly choose that exact number of craft sticks. Continue saying different numbers until all the craft sticks are gone.

Chenille-Stem Numbers

Prepare a set of 10 chenille stems for this activity. Thread large beads onto 10 chenille stems so that each stem has 1–10 beads. Once the set is made, mix up the chenille stems. Ask the student to quickly find and give you a certain stem. Randomly name every number value until all the chenille stems have been named and handed over.

Interlocking Cubes

Put together groups of interlocking cubes representing number values from 1 to 10. Using a screen to shield the cubes from the student's view, place one group of connected cubes behind the screen. Try using color patterns to make recognition easier. Then, remove the screen for 3–5 seconds, allowing the student to briefly view the set of cubes. Replace the screen and ask the student to determine the number of cubes. Repeat this activity 10–20 times in a daily session. Ensure that the student experiences a high rate of success.

More or Less?

Prepare a deck of playing cards by removing the face cards and cutting out all the numerals in the corners. The remaining cards display only a group of spades, hearts, diamonds, or clubs. Distribute two cards to the student, faceup. Have him or her compare the two numbers and quickly sort the cards into one of three piles: *more*, *less*, or *equal*. Continue to distribute two cards at a time, and have him or her sort the cards into the *more*, *less*, or *equal* piles. Play fast and have fun!

Skills and Activities
for Proficiency in
Grade 1

By focusing on deep understanding of the Essential Math Skills for Grade 1, we can allow many more students to find joy in the learning and practice of mathematics. Although this set of skills may appear simple enough, this learning year is among the most crucial to determining which students move ahead in school with a true appreciation for and love of mathematics.

The Grade 1 math skills should be learned to proficiency—a level of deep understanding that allows a student to use the skills in multiple contexts and in new situations. These five skills are not intended to be a comprehensive curriculum or an instructional program. First-grade mathematical learning experiences should include measuring, movement, prediction, basic addition and subtraction, and activities and projects that bring math to life. Use rich math vocabulary while engaging students in these activities (see *Math Vocabulary,* page 166).

Recommended Materials

- Abacus
- Classroom objects, (e.g., pencils, erasers, craft sticks, paper clips, buttons, beads, beans, interlocking cubes)
- Counters
- Base-ten units and rods
- Chenille stems with beads (see page 53)
- 10 empty egg cartons
- 10 plastic eggs
- Pennies
- Piggy bank
- Clothesline
- 100 metal washers
- Clothespins
- Toy hoop
- Pattern blocks
- Rubber stamps and ink pad
- Construction paper (4 colors)
- Stackable cups (4 colors)
- Stackable blocks (4 colors)
- Plastic clothespins (4 colors)
- Shoe box
- Keyboard
- Bingo chips or plastic discs
- Number cubes
- Interlocking cubes
- Spinner

Skills and Activities
for Proficiency in
Grade 1 *(cont.)*

Essential Math Skills

Skill 9: Counts objects with accuracy to 100

Skill 10: Replicates visual or movement patterns

Skill 11: Understands concepts of adding on or taking away (within 30), with manipulatives

Skill 12: Adds/subtracts single-digit problems on paper

Skill 13: Shows a group of objects by number (to 100)

Additional Resources

- *Hundred Chart* (page 150; hundredchart.pdf)
- *Thirty Chart* (thirtychart.pdf)
- *Here's What I Think* (page 148; whatthink.pdf) *(optional)*

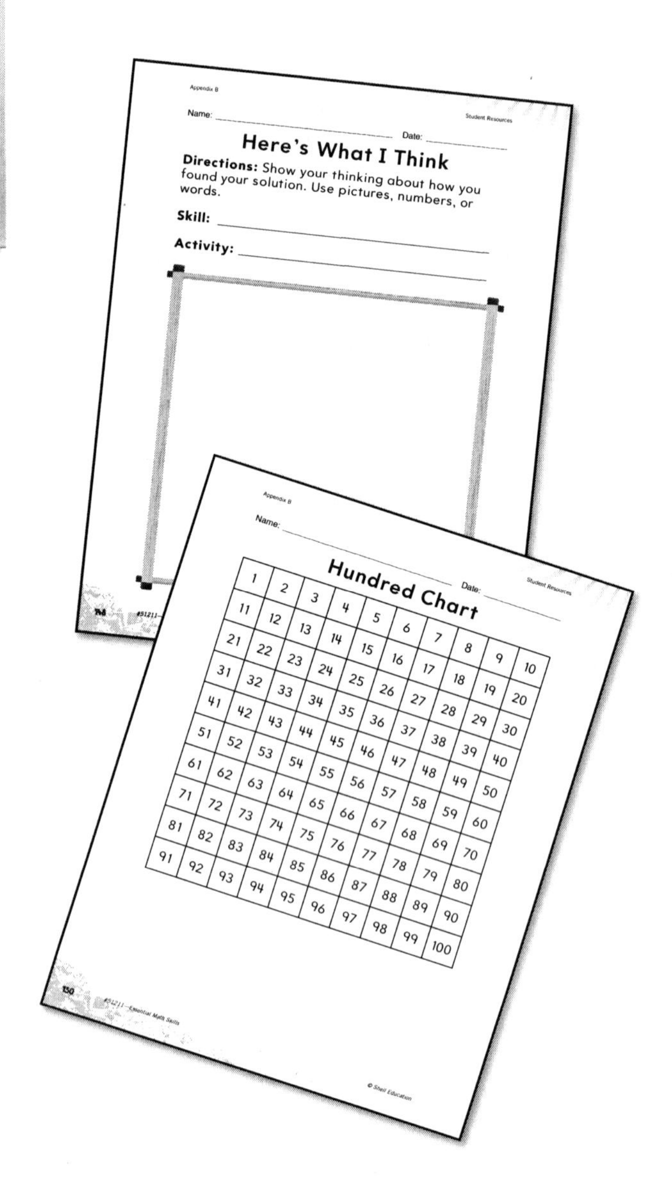

Name: Date:

Here's What I Think

Directions: Show your thinking about how you found your solution. Use pictures, numbers, or words.

Skill:

Activity:

Name: Date:

Hundred Chart

1	2	3	4	5	6	7	8	9	10
11	12	13	14	15	16	17	18	19	20
21	22	23	24	25	26	27	28	29	30
31	32	33	34	35	36	37	38	39	40
41	42	43	44	45	46	47	48	49	50
51	52	53	54	55	56	57	58	59	60
61	62	63	64	65	66	67	68	69	70
71	72	73	74	75	76	77	78	79	80
81	82	83	84	85	86	87	88	89	90
91	92	93	94	95	96	97	98	99	100

Grade 1 Skills

Skill 9

Counts objects with accuracy to 100

Skill Progression

Proficiency for this skill is demonstrated by the student's ability to swiftly count any number up to 100 with 100 percent accuracy on three or more days, using three or more different activities.

Intervention	Developing	Proficient
Counts fewer than 20 objects with accuracy	Sometimes counts objects with accuracy (20–100)	Consistently counts objects with accuracy (to 100)

Activities for Skill Development and Assessment

Absolutely Abacus: How Many Is This?

Beginning with number values that you are sure the student will recognize, move the beads on the abacus and ask, "How many is this?" Repeat this activity at least 10 times within a session. Gradually challenge the student until he or she can quickly count or recognize any number up to 100. You might display numbers in patterns (e.g., 22, 32, 42, 52), which the student will begin to recognize through practice and self-discovery. Avoid telling the student how to recognize these patterns; allow him or her to experience the learning that comes with discovery. Give plenty of practice to this activity so that the student can develop a deep understanding of the relative values of numbers up to 100.

Counting Fun

Using common classroom objects (e.g., pencils, interlocking cubes, bingo chips, buttons, beans, or paper clips), ask the student to count values up to 100.

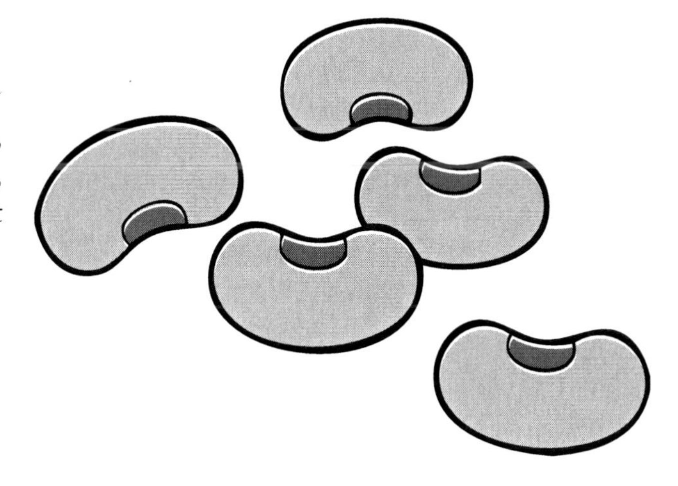

Hundred Chart

Using the *Hundred Chart* (page 150), ask students to place counters on the chart to show values up to 100.

Variation: Ask the student to place one counter on the chart at varying number positions. Try having him or her place a counter to show a number value and then move 10 spaces up or down the chart. Ask, "What number is the counter on now?"

Rods and Cubes

Use base-ten units and rods to show the student a number up to 100. Have him or her identify the number value.

Chenille-Stem Tens

Use chenille stems on which 10 beads have been placed, along with individual beads. Show the student a number value using the chenille-stem tens and ones (individual beads), and ask him or her to identify the number value.

Variation: Let the student show you the value of a number (to 100) using chenille-stem tens and individual beads.

Egg-O

Using egg cartons with two of the egg receptacles blocked or removed (leaving 10 receptacles in each carton), fill several cartons with 10 plastic eggs and one carton with fewer than 10. Have the student determine the total. If the student is familiar with the use of a ten frame, he or she will quickly connect this concept to the cartons of 10.

Piggy Bank Collection

Give the student a cup or a small bag of pennies, and ask him or her to count them out loud as he or she drops them into a piggy bank. Try collecting pennies for a charitable organization or a cause, and ask the student to count the total pennies each day until the donation goal is met.

Hang the Washers on the Line

Hang a clothesline on which 100 metal washers have been strung. Allow the student to separate the washers into groups of 10, using a clothespin to separate the groups. Then, ask him or her to identify various number values as quickly as possible.

Odds On *(Challenge Activity)*

On the *Fifty Chart* (page 152) or the *Hundred Chart* (page 150), have the student identify and color all the odd or even numbers.

Variation: Once the student has colored the odd and even numbers on a chart, teach the basics of skip counting. Count by two, starting with zero, or starting with one.

Grade 1 Skills

Skill 10

Replicates visual or movement patterns

Skill Progression

Proficiency for this skill is demonstrated by the student's ability to replicate four four-part patterns, with 100 percent accuracy on three or more days, using three or more different activities. The students should demonstrate this skill with both gross-motor and visual-motor patterns. Ask the student to follow oral directions for two- to four-step gross-motor patterns. Gradually increase difficulty, being aware that a student's listening memory can be a factor impacting success. While patterns are not addressed by the Common Core State Standards for Mathematics before Grade 3, learning to recognize and follow patterns requires the integration of both sides of the cortex, and is crucial for a general understanding of mathematics concepts.

Intervention	Developing	Proficient
Has difficulty replicating a two-step visual pattern (e.g., square, circle; red, blue) or a two-step movement pattern (e.g., clap hands, step forward)	Can sometimes replicate a two- or a three-step visual or movement pattern	Can consistently replicate three- and four-step visual and movement patterns

Activities for Skill Development and Assessment

Hoop Steps

Place a toy hoop on the floor and ask the student to watch what you do. Begin with a two-step sequence, such as *walk around the hoop once and then step inside it*. Have the student copy you. Continue with simple two-part sequences. As his or her skill improves, ask the student to copy more complicated three- and four-step gross-motor sequences.

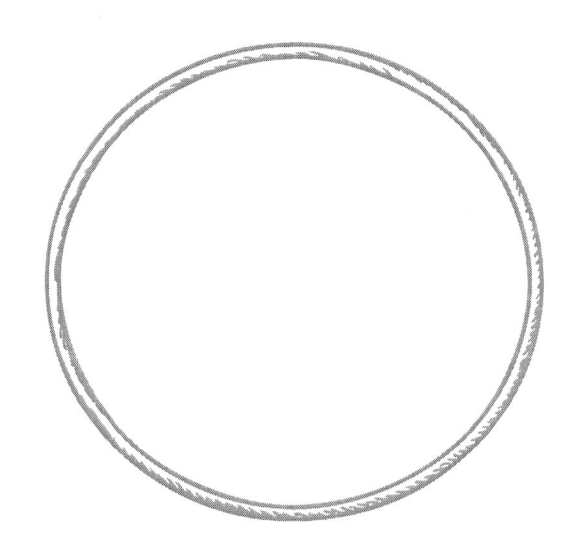

Two Step

Have the student stand next to you. Take two steps forward, and ask the student to copy you. Then, take two steps to the left, and have him or her copy the movement. Continue taking two steps in a variety of directions and asking the student to mirror them.

Once the student becomes successful following your lead, introduce a sequence of stepping patterns. You might model two steps forward, followed by two steps to the right. Gradually build up to a sequence of three or four directional changes, and after you model the movement, have the student copy it. He or she should end up standing by your side after each sequence. Don't make it too challenging, however, as you want to establish a high success rate and keep it within the student's ability.

Variation: Once the student has mastered a pattern, have him or her extend the pattern independently.

Hand Clap

Stand facing the student and begin doing the motions for and saying the words to a common rhyme or chant. Do it very slowly, one or two movements at a time. Tell the student to copy you after each set of movements you demonstrate. When he or she is ready, put them all together.

Animals in Action

Ask the student to help you create actions that represent different animals. For example, you might flap your arms like a bird, waddle like a penguin, or gallop like a horse. Have the student create a pattern, using two actions. As the student's skill improves, ask him or her to replicate more complicated three- and four-step gross-motor sequences (e.g., horse, penguin, bird).

Variation: Once the student has mastered a pattern, have him or her extend the pattern independently.

Follow the Leader

Begin by modeling two-part tasks, such as patting your head and then touching both knees. Once the student is successful with two-part tasks, try three-part gross-motor patterns such as stepping forward, then tapping both knees with your hands, and then clapping your hands over your head. Ask the student to copy you. As the student becomes skilled at copying three-part patterns, challenge him or her by adding a fourth task.

Motor Play

Identify gross-motor activities that the student can easily perform, possibly jumping jacks, skipping, galloping, marching, walking backward, hopping, or jumping from one foot to the other. Then, begin by modeling two-step patterns, such as taking a step backward and then doing a jumping jack. Gradually build to three- and four-part patterns.

Variation: Once the student has mastered a pattern, have him or her extend the pattern independently.

Pattern Blocks

Create a two-part, three-part, or four-part pattern, using pattern blocks. Have the student extend the pattern until it has been repeated three times.

Variation: Try creating patterns by tracing pattern blocks onto strips of paper. Then, have the student extend the pattern until it has been repeated three times.

Stamp in Sequence

Using rubber stamps, ink pads, and some strips of paper, begin a three-part or four-part pattern, using the stamps. Ask the student to continue the pattern you started.

Paper Chains

Using construction paper in at least four different colors, build a paper chain in a three-part or a four-part pattern. Have the student extend the pattern.

Tower of Power

Using a collection of large colored blocks, build a tower of stacked blocks in a three-part or a four-part color pattern. Ask the student to take over and continue the pattern as high as possible.

Variation 1: Gather cups of at least four different colors. Create a pattern of stacked cups, and have the student replicate and extend the pattern.

Variation 2: Ask the student to create a pattern, explain it to you, and then extend the pattern.

Patterns on the Line

Gather an assortment of at least four different-color plastic clothespins (or use permanent markers to color the ends of wooden clothespins). On a clothesline, create a color pattern with the clothespins and have the student extend the pattern.

Variation 1: Clip the clothespins to the edge of a shoebox, with the pattern extending around the sides of the box.

Variation 2: Ask the student to create a pattern, explain it to you, and then extend the pattern.

Musical Patterns

Using a keyboard, model a three- or four-step pattern and ask the student to extend the pattern. Experiment with patterns of varying complexity.

Absolutely Abacus: Bead Patterns

Create a pattern of grouped beads on the abacus (e.g., two beads, space, one bead, space, two beads, space, five beads). Have the student replicate the pattern on the next row of the abacus.

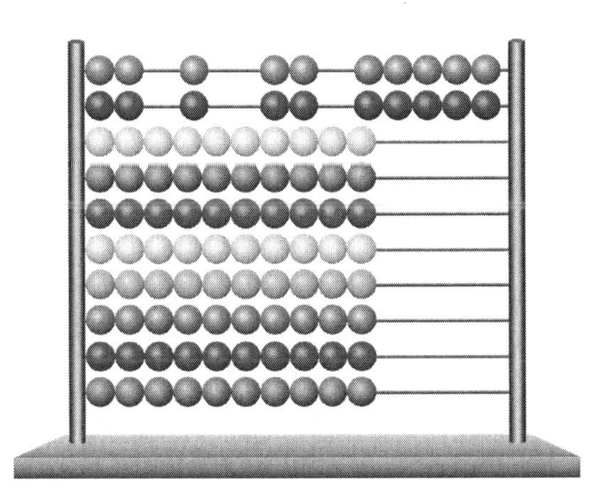

Grade 1 Skills

Skill 11

Understands concepts of adding on and taking away (within 30), with manipulatives

Skill Progression

Proficiency for this skill is demonstrated by the student's ability to quickly solve 10 addition or subtraction problems with a maximum sum or minuend of 30, using manipulatives, with 100 percent accuracy on three or more days, using three or more different activities.

Intervention	Developing	Proficient
Unable to add on or take away numbers from a group (within 10)	Uses manipulatives to add on or take away from a group, but must recount to find the total (within 30)	Using manipulatives, can add on or take away numbers and name the resulting number (within 30)

Activities for Skill Development and Assessment

Absolutely Abacus: What Number Is This?

For this activity, the quality of the student's learning will be dependent upon the use of numbers that offer a high success rate with a slight element of challenge to make it fun. The student should be finding correct responses at least 90 percent of the time. This skill helps set the stage for deep understanding of adding on and taking away.

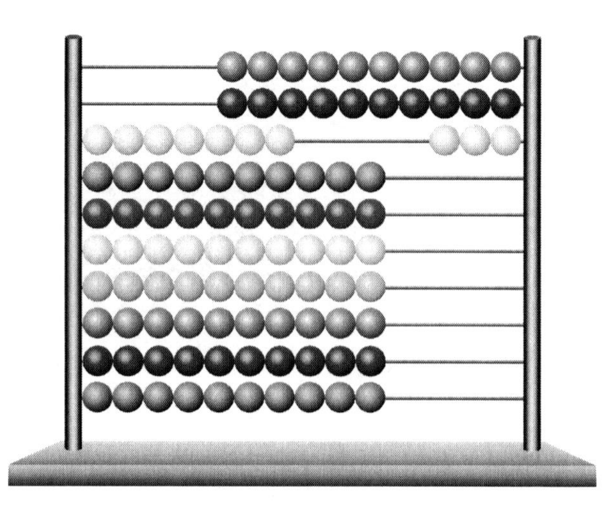

Beginning with number values that you are sure the student will recognize, move the beads on the abacus and ask, "What number is this?" This choice of words is deliberate. Your student is not just counting beads, steps, or marbles anymore. He or she is counting numbers! Gradually challenge the student until he or she can quickly count or recognize any number up to 30. Repeat this activity at least 10 times within a session.

Absolutely Abacus: Show Me

As in the last activity, begin with easy numbers to establish success. Gradually use larger numbers up to 30 (or greater, as appropriate). Ask the student to show you a certain number. Gradually challenge the student until he or she can quickly show you any number up to 30. Repeat this activity at least 10 times within a session.

Enjoy observing the student as he or she figures out new numbers. Sometimes the student will recognize groups and combinations. At other times, he or she many revert to one-by-one counting. Restrain your urge to point out patterns. The student will figure them out on his or her own, allowing for the complete understanding that will become part of his or her knowledge.

Absolutely Abacus: Show Me Plus

Ask the student to use the beads to show a certain number. Then, ask him or her to add a certain number more. Use addends less than 10. Start with small addends such as 2 or 3. Then, ask him or her to identify the total. Celebrate each success.

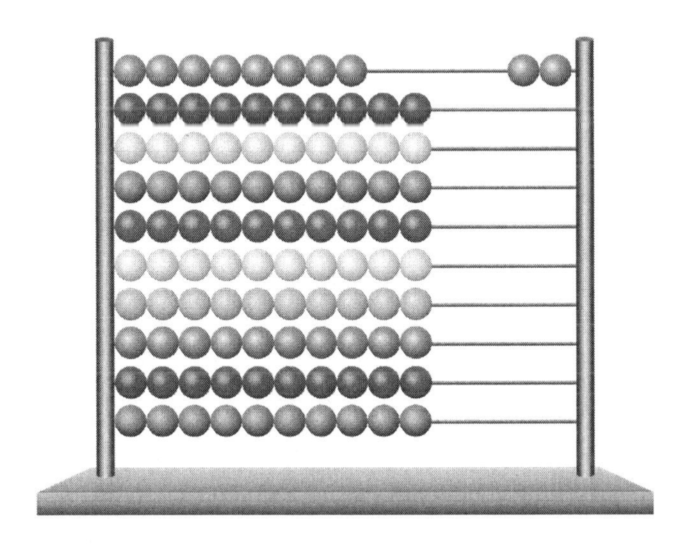

Experiment with patterns of numbers within the student's range of success. On one day, you may use 3 as the addend in most problems. On another day, you may always start with the number 15. On another day, you may use 5 as an addend in most problems. Repeat this activity at least 10 times in a session.

Don't rush through this skill. It is too important. Revisit this activity until the student can quickly identify and add any number combination up to 30. Take many weeks to practice this skill until the student has gained deep understanding.

Absolutely Abacus: Show Me Minus

Ask the student to use the beads to show you a certain number. Then, ask him or her to take away a certain number from that group and identify how many are left. Again, build success in this new activity by using smaller numbers and subtracting numbers less than 10. Be as concrete as possible, at first using the term *take away* in your instructions. Gradually introduce the use of *subtract* or *minus* into the instructions. Experiment with patterns of subtraction until the student can quickly subtract within 30. Have fun, and celebrate success.

Absolutely Abacus: Test the Teacher

In the spirit of fun (and motivated learning), allow the student to create some addition and subtraction problems for the teacher to complete on the abacus. Have the student make up 5–10 problems in a session.

Absolutely Abacus: Simple Story Problems

Make up simple yet interesting addition or subtraction story problems that the student can understand and solve, using numbers within 30. For example: *There are seven students on the bus. Three more hop on. Now how many students are on the bus? Show me on the abacus, and then tell me the solution.*

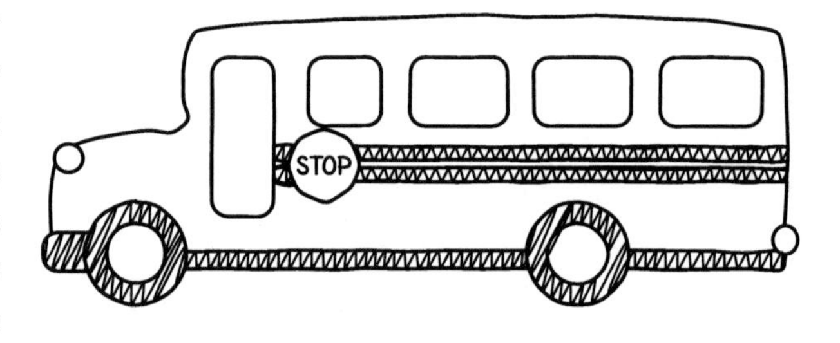

Ask the student to resist the urge to shout out the answer. Ask that they show the problem on the abacus before providing the solution. Practice at least five story problems in a session.

Whatever You've Got

Using bingo chips, pennies, buttons, beads, paper clips, erasers, craft sticks, interlocking cubes, or base-ten blocks, have the student solve addition and subtraction problems (within 30).

Chenille-Stem Sums

Use chenille stems on which 10 beads have been placed, and individual loose beads. Using sums within 30, show the student a number value using the chenille-stem tens and ones (beads), and ask him or her to add or subtract any number and name the sum or difference.

Thirty-Chart Fun

Distribute the *Thirty Chart* (thirtychart.pdf) to the student. Using bingo chips, discs, or pennies, ask the student to solve addition and subtraction problems by placing the manipulatives on the chart.

The People Mover

Begin with the entire class standing. Show how having students sit down (subtracting them from the large group) decreases the number of students standing. Then, solve addition and subtraction problems using the students as the math learning manipulatives. Have fun!

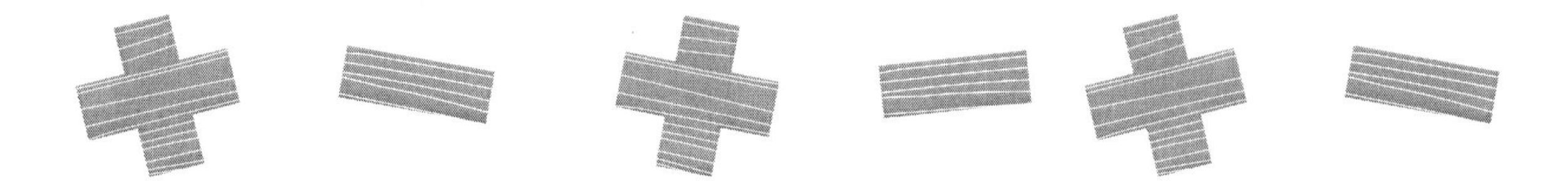

Grade 1 Skills

Skill 12

Adds/subtracts single-digit numbers on paper

Skill Progression

Proficiency for this skill is demonstrated by the student's ability to quickly add or subtract two single-digit numbers on paper with 100 percent accuracy on three or more days.

Intervention	Developing	Proficient
Needs assistance to add two single-digit numbers on paper	Adds two single-digit numbers on paper independently and with partial accuracy	Adds or subtracts two single-digit numbers on paper independently and accurately

Activities for Skill Development and Assessment

Absolutely Abacus: Problems on Paper

Pose addition or subtraction problems using single-digit numbers for the student to solve, using the abacus. Once the problem is correctly modeled on the abacus, ask the student to write the problem and solution on paper. Model several representations of the equation:

$$2 + 3 = 5$$

$$5 = 2 + 3$$

$$\begin{array}{r} 2 \\ +3 \\ \hline 5 \end{array}$$

This is the first time the student has been asked to use number symbols to represent his or her work on the abacus. Get excited when he or she does it correctly! Help this first connection with symbolic problem solving be joyous. Practice at least 5–10 problems in a session.

Roll and Record the Problem

Create a number-sentence frame for either addition or subtraction on a sheet of paper:

_______ + _______ = _______

_______ – _______ = _______

Have the student roll two number cubes, count the dots on each, and record each number in either of the first two blanks, as appropriate. Have the student determine the sum or difference, and record it in the blank following the equal sign.

Variation: Have the student roll three or more number cubes and find the sum or difference.

Interlocking Cubes or Base-Ten Blocks Story Problems

Using interlocking cubes or base-ten blocks, first ask the student to solve single-digit addition and subtraction problems using the manipulatives, and then write the problem and solution on paper.

Variation: Use three or more numbers in an addition or a subtraction problem.

On Paper Only

Using single-digit numbers, give the student a set of 10 addition or subtraction problems on paper. Allow him or her to look at the abacus or other manipulatives but not touch them. Ask the student to solve the 10 problems. Do not rush through this activity or put any time limits on it. Construct problems systematically (e.g., use a consistent addend) or randomly, as appropriate for the learning needs of the student.

Write What You Hear

Dictate one addition or subtraction problem, using single-digit numbers, and have the student write it on paper. Ask him or her to solve the problem. Allow the student to use the abacus or draw a picture to support the problem-solving process, if needed. As the student demonstrates competence at listening, writing the problem, and then solving the problem, gradually increase the number of problems given in one learning session (up to 10 per session). Do not time the student or hurry through this activity. Offer problems that follow a pattern or select problems randomly, as appropriate for the learning needs of the student.

Spin the Wheel

Using a spinner (spinner.pdf), ask the student to construct his or her own addition and subtraction problems and write them on paper. Spin the wheel twice to pick two numbers for addition. For subtraction, spin the wheel twice and subtract the smaller number from the larger number.

Variation: Allow the student to spin three to four single-digit numbers, write them as an addition problem on paper, and then solve the problem.

Shows a group of objects by number (to 100)

Skill Progression

Proficiency for this skill is demonstrated by the student's ability to show any number value up to 100 without counting each object, with 100 percent accuracy on three or more days, using three or more different activities.

Intervention	Developing	Proficient
Shows a group of objects of fewer than 25, using manipulatives	Shows a group of objects to 50, using manipulatives	Shows a group of objects to 100, using manipulatives

Activities for Skill Development and Assessment

Absolutely Abacus: Show Me (to 100)

Begin with number values that you are sure the student can successfully show on the abacus, and ask him or her to show you a certain number. Begin with a number under 30, and then ask for numbers following a pattern, such as 25, 32, 39, 46; or 16, 26, 36, 46. Practice this activity for many weeks until there is no question that the student understands the value of every number up to 100 and can quickly show any number value on the abacus. Repeat this activity at least 10 times within a session.

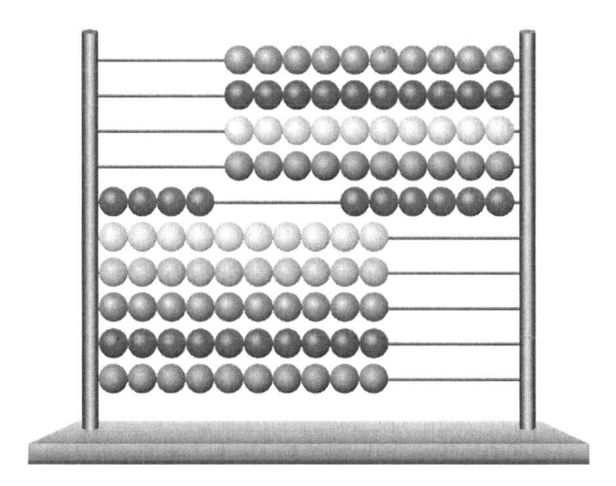

Variation: Ask the student to show a number value on the abacus and then draw an array to show the number of tens and ones.

Skill 13

Common Classroom Manipulatives

Using common classroom objects (e.g., pencils, interlocking cubes, paper clips), ask the student to show values up to 100. Allow the student to figure out ways to organize objects into groups of 10 (e.g., using rubber bands to group pencils or snapping cubes together in towers of 10), which facilitates modeling larger numbers.

Show It on the Hundred Chart

Say the name of any number under 100. Ask the student to use a manipulative (e.g., paper clips or interlocking cubes) to show that quantity. Then, have the student place a counter on that number on the *Hundred Chart* (page 150).

Base-Ten Number Models

Ask the student to show number values to 100 using a combination of ones (base-ten units) and tens (base-ten rods). Repeat this activity at least 10 times within a session.

Variation: Show the number value using units and rods and then show the same value on an abacus or by drawing an array.

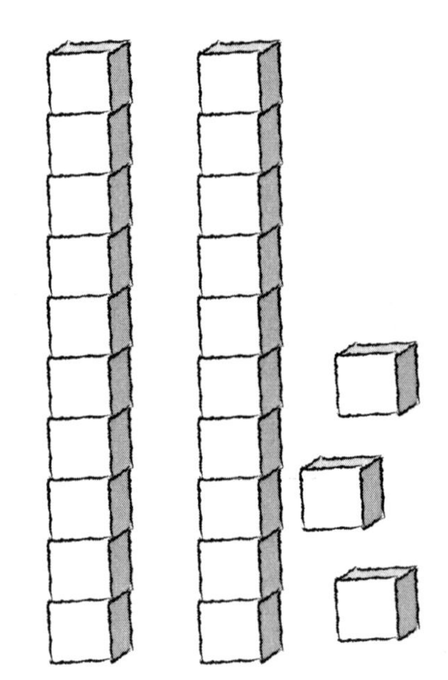

Chenille Stems to 100

Use chenille stems on which 10 beads have been placed, and individual loose beads. Ask the student to show number values to 100 using a combination of ones (beads) and tens (stems). Repeat this activity at least 10 times within a session.

Variation: Show the number value using stems and beads. Then, show the same value on an abacus or by drawing an array.

Egg-O Fill Up

Using egg cartons with two of the egg receptacles blocked or removed (leaving 10 in each carton), ask the student to show number values to 100 by filling the empty cartons with plastic eggs to represent tens and counters to represent ones.

Variation: Show the number value using the egg carton, and then show the same value on an abacus or by drawing an array.

Move the Washers on the Line

Hang a clothesline on which 100 metal washers have been strung. Ask the student to show you various number values as quickly as possible. Allow him or her to use clothespins to create groups of 10, which facilitates showing the value of larger numbers.

Variation: Show the number value using the metal washers, and then show the same value on an abacus or by drawing an array.

Skills and Activities

for Proficiency in Grade 2

The essential skills provide a guide to a set of skills that are indispensable, and for these learning outcomes, only proficiency is an acceptable standard. Take all the time necessary to help the student completely master these skills and use them within a variety of contexts. Help students truly understand these mathematical concepts in a way that allows them to love math for life.

Remember, the activities in this book are not intended to be a complete math program. These activities should be considered a supplement to the projects and activities typically found in any quality mathematics curriculum. Deep understanding and application of the second-grade essential math skills gives students the foundation to have a lifelong appreciation of math. Use rich math vocabulary while engaging students in these activities (see *Math Vocabulary,* page 166).

Recommended Materials

- Abacus
- Counters
- Base-ten units and rods
- Chenille stems with beads
- Clothesline
- Metal washers
- Clothespins
- Chocolate chips
- Egg cartons or muffin pans
- Small objects (e.g., buttons, beads, marbles, plastic chips, stones)
- Tagboard or construction paper
- Play money
- Paper plates
- Tape or chalk
- Beanbags
- Index cards
- Coffee mugs
- Rubber bands
- Craft sticks
- Sun visors (4)

Skills and Activities

for Proficiency in

Grade 2 *(cont.)*

Essential Math Skills

Skill 14: Quickly recognizes number groups (to 100)

Skill 15: Adds to/subtracts from a group of objects (within 100)

Skill 16: Adds/subtracts two-digit numbers on paper

Skill 17: Counts by 2, 3, 4, 5, and 10, using manipulatives

Skill 18: Solves written and oral story problems using the correct operations *(addition and subtraction)*

Skill 19: Understands/identifies place value to 1,000

Additional Resources

- *Hundred Chart* (page 150; hundredchart.pdf)
- *Blank Hundred Chart* (page 151; blankhundred.pdf)
- *Picture Cards* (picturecards.pdf)
- *Place-Value Chart* (page 149; placevaluechart.pdf)
- *Place-Value Cards* (placevaluecards.pdf)
- *Skill 18 Story Problems* (skill18problems.pdf)
- *Here's What I Think* (page 148; whatthink.pdf)

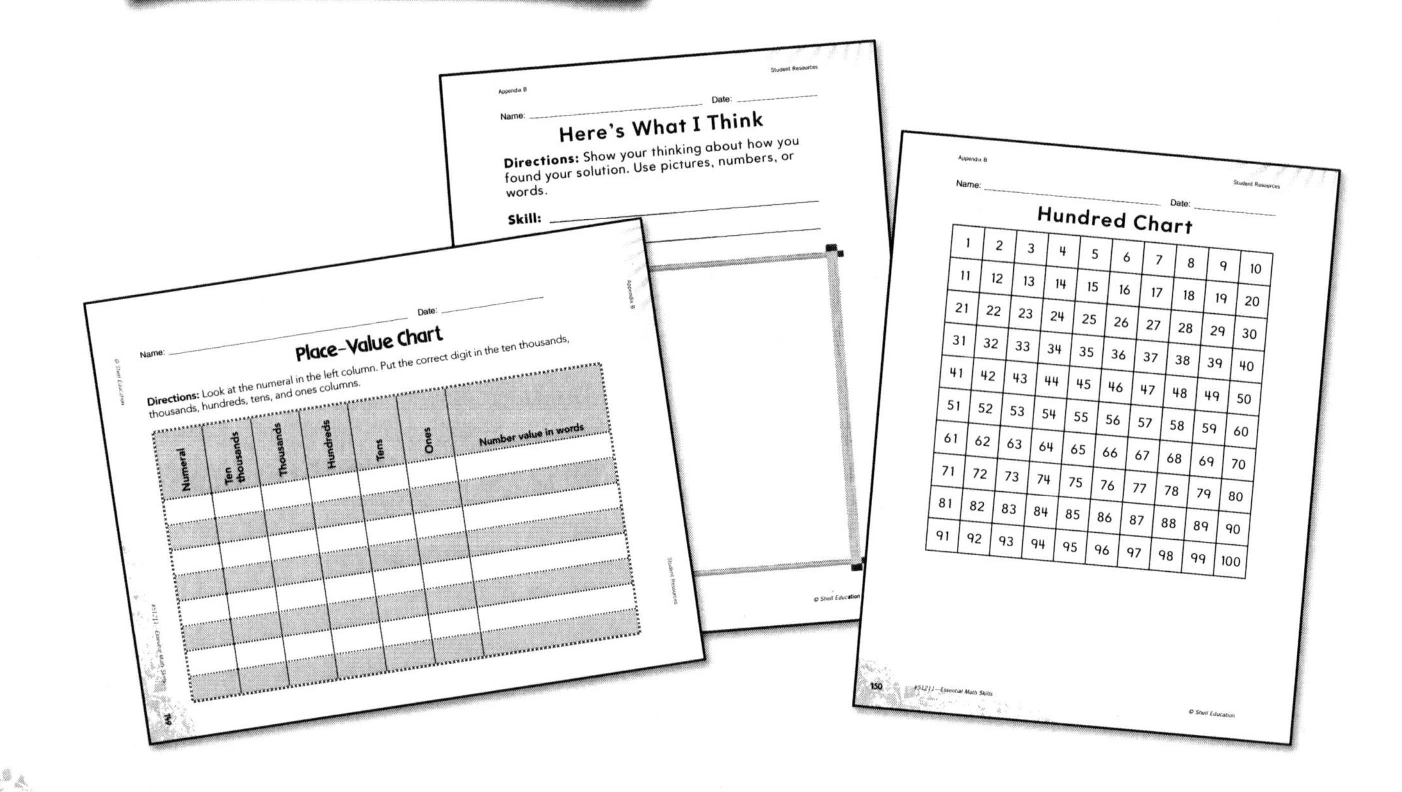

Grade 2 Skills

Skill 14

Quickly recognizes groups of objects (to 100)

Skill Progression

Proficiency for this skill is demonstrated by the student's ability to name any 10 number values up to 100, each within five seconds, with 100 percent accuracy on three or more days, using three or more different activities.

Intervention	Developing	Proficient
Using manipulatives, recognizes number groups of fewer than 25	Using manipulatives, recognizes number groups of 25–75	Using manipulatives, quickly recognizes number groups to 100

Activities for Skill Development and Assessment

Absolutely Abacus: Name That Number

Beginning with number values that you are sure the student can quickly name, move the beads and ask, "What number is this?" Repeat this activity at least 10 times within a session. Provide plenty of practice with this activity so that the student can develop a deep understanding of the relative values of numbers up to 100. Gradually challenge the student until he or she can recognize any group of beads up to 100 within five seconds.

Find It on the Hundred Chart

Using the *Hundred Chart* (page 150), place one counter on the chart, indicating a number. Ask the student to quickly name that number.

Variation 1: Ask the student to place counters on each square of the chart showing a number value up to 100. Then, ask the student to name the value.

Variation 2: Ask the student to place one counter on the chart at varying number positions. Use the *Blank Hundred Chart* (page 151) for an additional challenge.

Variation 3: Place a counter on the chart and then ask the student to quickly name the value that is 10 more or 10 less.

Rods and Cubes

Use base-ten units and rods to show the student a number value up to 100. Have the student quickly identify the number value. Repeat this activity at least 10 times within a session.

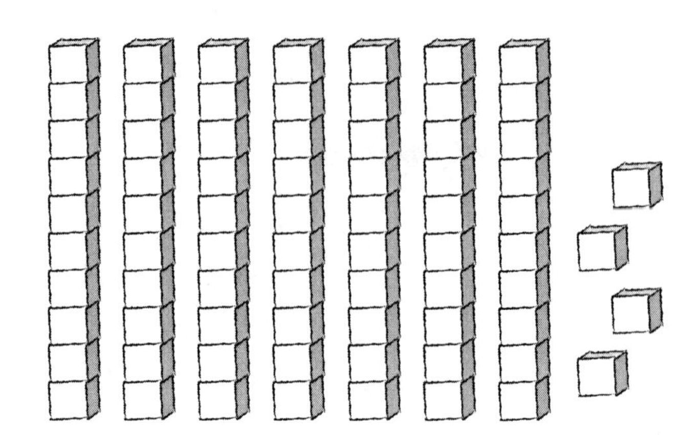

Variation 1: Name a number value, and ask the student to quickly show the value, using base-ten units and rods.

Variation 2: Have the student identify the value of a group of units and rods and then draw an array showing this value.

Ten Stems

Using chenille stems on which 10 beads have been placed and also separate individual beads, ask the student to identify number values to 100, using a combination of ones and tens. Repeat this activity at least 10 times within a session.

Variation 1: Name a number value, and ask the student to quickly show the value using ones and tens.

Variation 2: Have the student identify the value of a group of stems and beads and then draw an array showing this value.

On the Line

Hang a clothesline on which 100 metal washers have been strung. Show the student a number value, using clothespins to separate groups of 10. Ask him or her to quickly name that value.

Activities to Extend Deep Understanding

More or Less?

Have the student identify the value of a group on the abacus. Then, ask him or her to name the value that is 10 more or 10 less.

Variation: Ask the student to identify the value of a group on the abacus. Then, ask him or her to name the value that is 20, 30, 40, or 50 more or less.

Hundred Closure

Have the student identify the value of a number on the abacus or on the *Hundred Chart* (page 150). Then, ask him or her to name the additional amount needed to reach 100.

Test the Teacher

Allow the student to show a number value using the abacus, *Hundred Chart* (page 150), base-ten blocks, stems and beads, or washers on a string, and have the teacher quickly identify the value of this group.

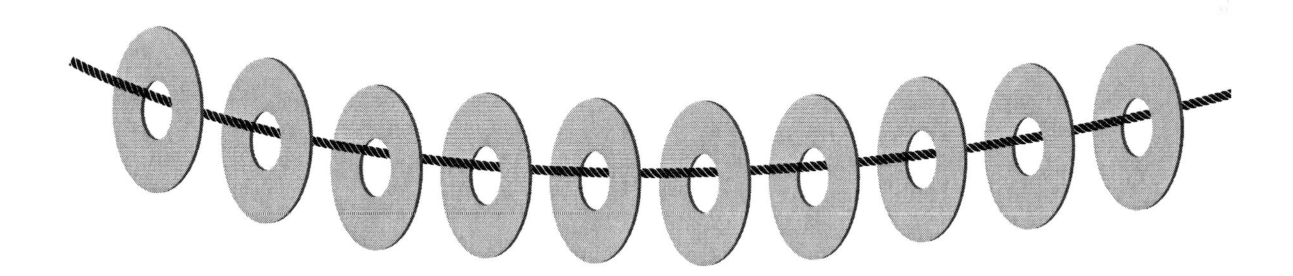

Grade 2 Skills

Skill 15

Adds to/subtracts from a group of objects (within 100)

Skill Progression

Proficiency for this skill is demonstrated by the student's ability to add or subtract number values within 100 within 10 seconds with 100 percent accuracy on three or more days, using three or more different activities.

Intervention	Developing	Proficient
Needs assistance to add to a group of objects and recognize the sum	Accurately adds to a group of objects but needs assistance with subtraction	Accurately adds to and subtracts from a group of objects within 100

Activities for Skill Development and Assessment

Absolutely Abacus: Show Me Plus

Ask the student to show a certain number on the abacus. Once completed, ask him or her to add a certain amount more. Then, ask the student to identify the total. Begin with numbers that allow for a high rate of success. Gradually increase the difficulty until he or she can show you a number and then add on, finding any sum within 100.

Experiment with patterns of numbers within the student's range of success. On one day, you may focus on 10 as an addend. On another day, you may always start with the number 15. On another day, you may frequently use 5 as an addend. On another day, 23 might be the addend of focus.

Repeat this activity at least 10 times in a session. Don't rush through this skill. Continue enjoying this activity until the student can confidently and quickly add any number combination up to 100.

Variation: Give the student a set of addition problems on paper. Ask him or her to solve each problem on the abacus and then record the answer on paper.

Absolutely Abacus: Two-Digit Addition with Regrouping

Give the student a set of two-digit addition problems that require regrouping, such as 44 + 17, 12 + 58, or 16 + 26. Ask the student to solve each problem on the abacus and write the solution on paper. Allow the student to solve at least 10 problems that require regrouping within one session. When using the abacus, the student may not realize that he or she is regrouping, but this practice will provide the foundation for understanding the concept of regrouping before moving to the abstract representation on paper.

Addition: From Abacus to Drawing

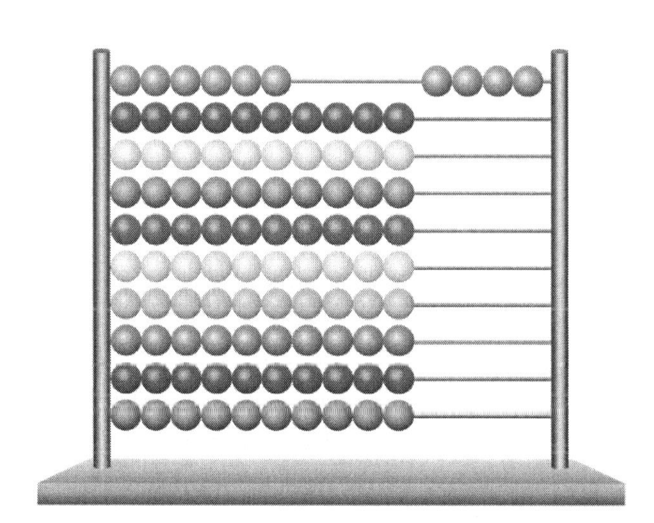

Give a set of addition problems to the student. Ask him or her to solve each problem on the abacus, record the solution, and then draw the problem and the solution.

Absolutely Abacus: Show Me Minus

Ask the student to show a certain number. Once completed, ask him or her to take away a certain number of beads. What's left? Begin with numbers and patterns that allow for a high rate of success. Gradually increase the difficulty until he or she can show you any number (to 100), then take away any value, and quickly and confidently find the solution. Make the activity fun. Give the student the time necessary to develop deep understanding and celebrate success.

Variation: Give the student a set of subtraction problems on paper. Ask him or her to solve each problem on the abacus and then record the solution on paper.

Absolutely Abacus: Two-Digit Subtraction with Regrouping

Give the student a set of two digit subtraction problems that require regrouping, such as 44 – 17, 58 – 19, or 61 – 2. Ask the student to solve each problem on the abacus, and then record the solution on paper. Have him or her solve at least 10 problems using regrouping in one session. When using the abacus, the student may not realize that he or she is regrouping, but this practice will provide the foundation for understanding the concept of regrouping before moving to the abstract representation on paper.

Subtraction: From Abacus to Drawing

Give a set of subtraction problems to your student. Ask him or her to solve each problem on the abacus, record the answer on paper, and then draw an array showing the problem and the solution.

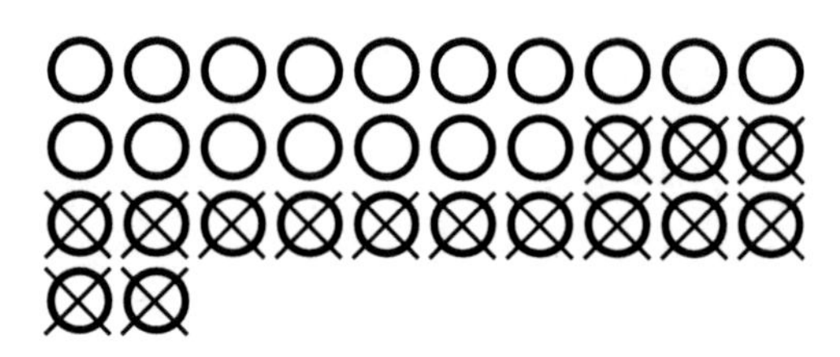

Absolutely Abacus: Test the Teacher

In the spirit of fun (and good learning), allow the student to make up some addition and subtraction problems on the abacus for the teacher. Have the student make up 5–10 problems in a session.

Hundred Chart Sums and Differences

Distribute the *Hundred Chart* (page 150). Ask the student to place one counter on any number. Then, ask him or her to add or subtract a certain number from that position and name the sum or difference. Practice using patterns of numbers within the student's range of success. Gradually increase the difficulty until he or she can quickly name a number, add or subtract from that number, and name the solution (within 100).

Rods and Cubes

Using base-ten units and rods, ask the student to show you any number value up to 100, add or subtract a number value, and name the sum or difference. As always, practice with patterns and allow plenty of time to develop deep understanding.

Variation: Ask the student to solve a problem using rods and cubes and then record the problem on paper.

Stem Tens

Use chenille stems on which 10 beads have been placed and also individual beads. Ask the student to show you any number value up to 100, add or subtract a number value, and name the sum or difference.

Variation: Ask the student to solve a problem using stem tens and beads and then record the problem on paper.

Washer Addition and Subtraction

Hang a clothesline on which 100 metal washers have been strung. Ask the student to show you any number value up to 100, using clothespins to separate the washers into groups of 10. Then, have the student add or subtract a number value and name the sum or difference. Practice using patterns, and allow plenty of time for the student to deeply understand addition and subtraction of numbers within 100.

Variation: Ask the student to solve a problem using washers, and then record the problem on paper.

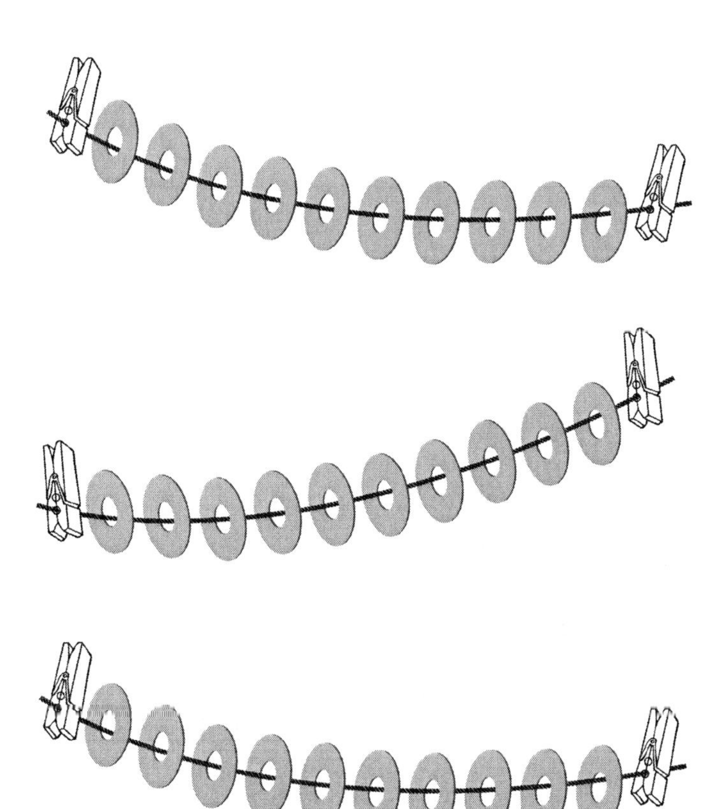

Adds/subtracts two-digit numbers on paper

Skill Progression

Proficiency for this skill is demonstrated by the student's ability to quickly solve 10 addition or subtraction problems on paper, using any combination of two-digit numbers (with solutions within 100) with 100 percent accuracy on three or more days.

Intervention	Developing	Proficient
Needs assistance to add two-digit numbers on paper	Accurately adds two-digit numbers on paper but needs assistance with subtraction	Accurately adds and subtracts two-digit numbers on paper

Activities for Skill Development and Assessment

Absolutely Abacus: Problems on Paper

Ask the student to solve an addition and subtraction problem, using one-digit and two-digit numbers within the problem on the abacus or counting frame, as before. Once the problem is correctly solved on the abacus, ask him or her to write the problem and solution on paper.

Written problems should look like this:

$$\begin{array}{r} 33 \\ \underline{+\,17} \\ 50 \end{array} \qquad \begin{array}{r} 33 \\ \underline{-\,17} \\ 16 \end{array}$$

The development of deep understanding of two-digit addition and subtraction takes a lot of practice and should not be rushed. Enjoy the sessions, making them just challenging enough for the student to give full attention while successfully completing almost all the problems. Practice at least 5–10 problems in a session.

Practice Sets Using the Abacus

Give the student a set of addition problems on paper. Ask him or her to solve each problem on the abacus and then on paper.

Absolutely Abacus: Two-Digit Addition with Regrouping

Give the student a set of two-digit addition problems that require regrouping, for example:

$$\begin{array}{r} 44 \\ +\ 17 \\ \hline \end{array} \qquad \begin{array}{r} 12 \\ +\ 58 \\ \hline \end{array} \qquad \begin{array}{r} 16 \\ +\ 26 \\ \hline \end{array}$$

Make the abacus available for use, but encourage the student to solve the problems without it. Have him or her solve at least 10 problems using regrouping in one session.

Addition: From Abacus to Paper to Array

Give a set of addition problems to the student. Ask him or her to solve each problem on the abacus, then on paper, and then again by drawing an array for the problem and the solution.

Absolutely Abacus: Two-Digit Subtraction with Regrouping

Give the student a set of two-digit subtraction problems that require regrouping, for example:

$$\begin{array}{r} 44 \\ -\ 17 \\ \hline \end{array} \qquad \begin{array}{r} 58 \\ -\ 19 \\ \hline \end{array} \qquad \begin{array}{r} 61 \\ -\ 22 \\ \hline \end{array}$$

Make the abacus available for use, but encourage the student to solve the problems without it. Have him or her solve at least 10 problems using regrouping in one session.

Subtraction: From Abacus to Paper to Array

Give a set of subtraction problems to the student. Ask him or her to solve each problem on the abacus, record the answer on paper, and then draw an array showing the problem and the solution.

Rods and Units Story Problems

Using base-ten rods and units, first ask the student to solve addition and subtraction problems (within 100) using the manipulatives, and then write the problem and solution on paper.

Variation: Have the student solve the problem using the *Hundred Chart* (page 150), chenille stems and beads, or washers on a clothesline. Then, have the student solve it on paper.

On Paper

Using two-digit numbers, give the student a set of 10 addition or subtraction problems on paper. Do not time the student or hurry through this activity. Construct problems using patterns, or randomly, as appropriate for the learning needs of the student. Allow the student to use the abacus if needed to understand the problem, and gradually remove the manipulative support as appropriate. This activity can be used in conjunction with problems found in any mathematics text or activity book.

Money Honey

Present two-digit addition and subtraction story problems using money after teaching the student the symbols for cents (¢) and dollars ($).

Counts by 2, 3, 4, 5, and 10, using manipulatives

Skill Progression

Proficiency for this skill is demonstrated by the student's ability to quickly skip-count by 2, 3, 4, 5, and 10, with 100 percent accuracy on three or more days, using three or more different activities.

Intervention	Developing	Proficient
Unable to skip-count using manipulatives	Consistently counts by 2 and 5, using manipulatives	Consistently counts by 2, 3, 4, 5, and 10, using manipulatives

Activities for Skill Development and Assessment

Absolutely Abacus: Skip Counting

Using the abacus, ask the student to count by a certain number (2, 3, 4, 5, or 10) to a multiple of 10. For example, ask the student to count by 3 to 30. First, he or she moves three beads and then says *three*. Then, he or she moves another set of three beads and now says *six*. Every time the student says a number, that number should be clearly shown on the abacus, using every bead in the top row before moving down to the next row. Practice several number sequences per day.

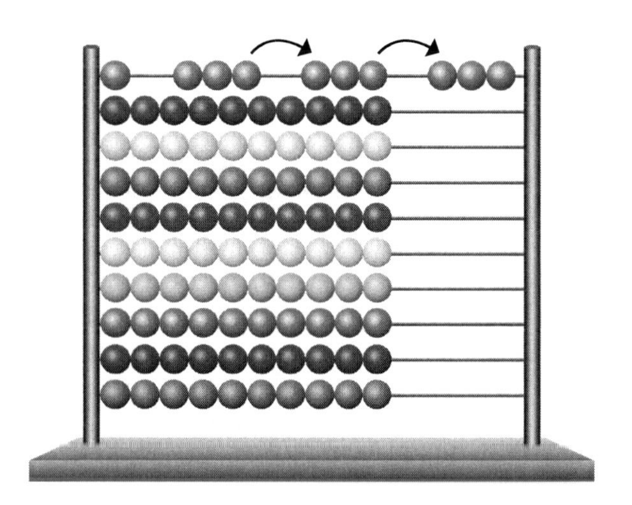

Hands Up, Hands Down

Gather a group of five students in a circle. Have them raise their hands one at a time while going twice around the circle, counting by 5 to 50. Then, try counting backward from 50 to 0 by having the students put their hands down one at a time. Ask them to tuck their thumbs and count by 4 to 40. Then, tuck two fingers and count by 3 to 30.

String Theory

Hang a clothesline on which 100 metal washers have been strung. Ask the student to count by 2, 3, 4, 5, or 10, separating each group with a clothespin.

Chocolate Counts

Pour a bag of chocolate chips or chocolate candies onto a paper plate. Ask the student to count the group by a certain number (e.g., 2, 3, 4, 5, or 10), and then set aside any leftovers. Then, pick a different number to use for skip counting and continue to set aside the leftovers. You may choose to allow the student to eat the leftover chocolate after each round of skip counting, if desired.

Carton Counting

Using empty egg cartons or muffin pans, count out a number of objects (e.g., buttons, beads, marbles, plastic chips, or stones) for each space, and then have the student skip-count by that number to find the total.

Nickeled and Dimed

Use nickels to practice skip counting by 5 and dimes to practice skip counting by 10.

Variation: Make up story problems using money. For example: *You want to buy a toy that costs 95 cents. If you use only nickels to buy it, how many nickels will you need?*

Get in Line

Use vinyl spots, tape, or chalk to mark a line of 10 spots on the floor. Ask the student to step on the spots and skip-count by a certain number with each step. If working with a small group of students, provide each student with a number card (2, 3, 4, 5, or 10), and then ask each student to walk the path while counting by that number. When finished, ask the students to swap cards and repeat the activity with a new number.

Hands

Trace your hand on tagboard or construction paper, and make 10 cutouts. Give the student the 10 cutout hands, and then ask him or her to count by 5 to a number up to 50. While counting by 5, he or she should lay down the hands in a row. Then, have the student try counting backward.

Variation: Create sets of 10 cutout hands showing 4, 3, or 2 fingers. Use the sets to have the student practice counting by 4 to 40, 3 to 30, and 2 to 20.

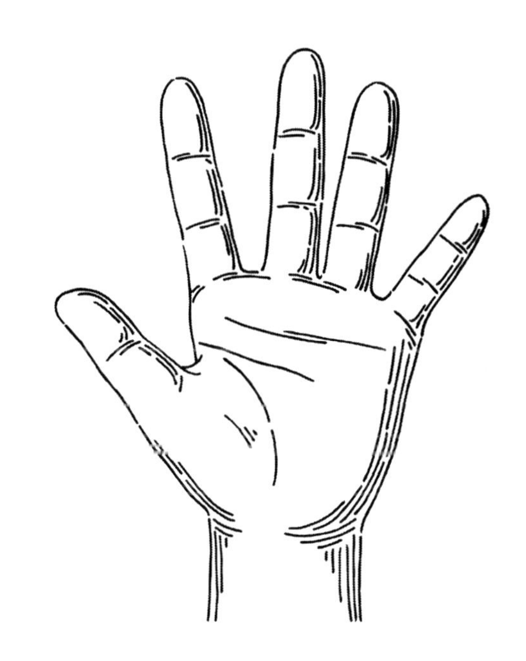

Picture Counting

Make 10 copies of the *Picture Cards* (picturecards.pdf) to have the student practice counting by a certain number (e.g., the bike has 2 wheels, a traffic signal has 3 lights, a horse has 4 legs, a sea star has 5 arms). Have the student count by the appropriate number up to its multiple of 10.

Pass the Plates

Give the student 10 small paper plates on which three symbols (e.g., circles, stars, hearts, diamonds, or squares) have been drawn. Ask him or her to count by 3 and pass you a plate with each count. Try the same activity with 2, 4, 5, or 10 symbols on each plate.

Absolutely Abacus: Up-and-Down Skip Counting

Using the abacus, ask the student to count by 2, 3, 4, 5, or 10 to a multiple of 10 (e.g., by 3 to 30). Provide plenty of practice time for this activity, which helps establish an understanding of multiples. Once the student is skilled at counting up, introduce counting backward to zero, showing the correct set of beads as you say each number.

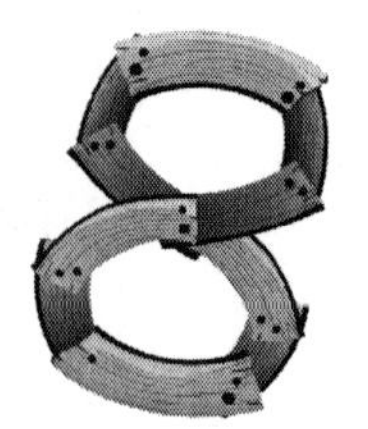

Variation: Have the student count to a number other than its multiple of 10. For example, count by 4 to 16, by 5 to 30, or by 10 to 70. This activity deserves a lot of practice time!

Skip-Counting Catch

When the student has mastered skip counting by any base number, consider a catching game to improve speed and automaticity. Using a beanbag, play underhand catch with the student. Each time he or she catches the beanbag, the student must say the next number in the skip-counting sequence.

Grade 2 Skills

Skill 18

Solves written and oral story problems using the correct operations *(addition and subtraction)*

Skill Progression

Proficiency for this skill is demonstrated by the student's ability to solve 10 written or oral story problems using addition or subtraction with numbers up to 100 with 100 percent accuracy on three or more days. In addition to using the provided activities, students need to receive instruction on the problem-solving process.

Intervention	Developing	Proficient
Unable to solve written and/or oral story problems with guidance	Able to solve written and oral story problems with guidance	Consistently able to solve written and oral story problems by developing a plan, solving problems using correct operations, and evaluating the solution

Activities for Skill Development and Assessment

Take It to the Story-Problem Bank

Ask the students to write story problems on index cards. Use these to make a bank of story problems that can be used for practice.

Absolutely Abacus: Simple Oral Story Problems

Make up simple but interesting story problems using addition or subtraction that the student can understand and solve, using numbers up to 100. Ask him or her to solve the problem first on the abacus and then on paper. As the student's skills develop, remove the abacus.

Sample Problems

There are 17 students on the bus. Five more hop on. Now, how many students are on the bus? Show me on the abacus, and then tell me the solution.

In the cupboard, there are 30 apples. While the students are out at recess, 4 apples are mysteriously eaten. How many are left?

There are 28 students in Ms. Loeffler's class. There are 30 students in Mr. Armstrong's class. They are going on a field trip. How many students will be on the bus? If the bus has 60 seats, how many seats will be empty?

Ask the student to resist the urge to shout out the answer, and encourage him or her to show the mathematics on the abacus before providing an answer. Practice at least five story problems in a session.

As the student becomes skilled at this process, he or she might say, "I don't need the beads" or "I already know the answer" or "I can see the abacus in my head." This indicates that the student is developing a deep understanding of basic number concepts.

Activities to Extend Deep Understanding

Absolutely Abacus: Half the Class

Have half the class write a set of five story problems, using numbers up to 100. Then, ask the other half to solve the problems first on the abacus and then on paper. As students' skills develop, remove the abacus.

Absolutely Abacus: Written Story Problems

Develop a library of story problems using addition or subtraction with numbers up to 100. Choose 5–10 problems at the readiness level of each individual student. Ask the student to solve a problem first on the abacus and then on paper. As the student's skills develop, remove the abacus. Try using multistep story problems with students who are ready for that challenge, ensuring that they are at least 90 percent successful with effort.

Easiest to Most Difficult

Give the student 10 randomly chosen story problem cards, and ask him or her to order them from easiest to most difficult. Ask him or her to explain the choice of order.

Oral Tradition

Give interesting story problems to the student using oral language only. This involves listening skills as well as math skills. Teach the student techniques for jotting down the relevant numbers and then accurately choosing the correct operation. Experiment with multistep story problems.

Variation: Ask the student to explain his or her choice of operation and the solution to the problem.

Story Problem Practice

Distribute *Skill 18 Story Problems* (skill18problems.pdf) to the student for additional practice with grade-level-appropriate story problems.

Grade 2 Skills

Skill 19

Understands/identifies place value to 1,000

Skill Progression

Proficiency for this skill is demonstrated by the student's ability to sort numerals into place value (to thousands) with 100 percent accuracy on three or more days, using three or more activities.

Intervention	Developing	Proficient
Does not yet understand place value for the ones, tens, hundreds, or thousands place	Consistently identifies place value for the ones and tens place. Sometimes identifies place value for hundreds and/ or thousands place	Consistently understands and identifies place value for the ones, tens, hundreds, and thousands place

Activities for Skill Development and Assessment

Place-Value Sticks

You will need nine coffee mugs, a box of rubber bands, and 999 craft sticks. Put rubber bands around groups of 10 craft sticks. Place 10 banded groups in each mug. You will have nine banded groups of 10 and nine individual sticks left.

Ask the student to show you the value of a three-digit number by arranging mugs, groups of 10, and ones from left to right. Then, ask him or her to write the value on paper. Repeat this activity 5–10 times in one learning session.

Roll 'Em

Give the student three or four number cubes. Have him or her roll the number cubes and create the largest number possible with the values rolled and then say the number. Repeat the activity several times.

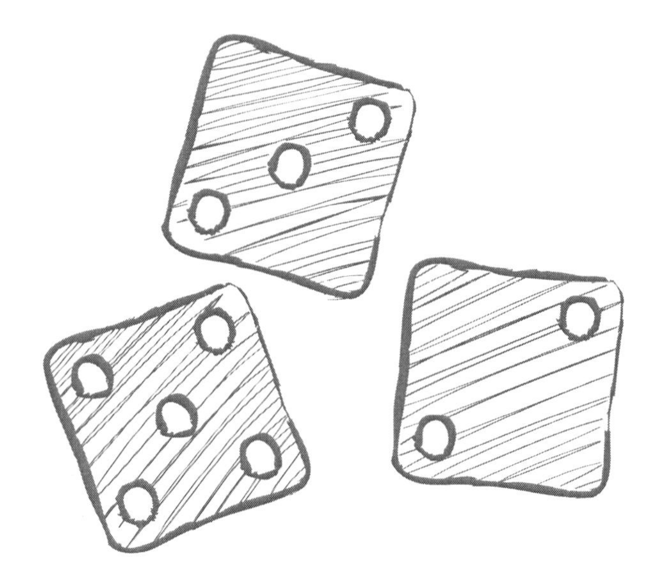

Place-Value Chart

Using the *Place-Value Chart* (page 149), give the student a set of two-, three-, or four-digit numbers, and ask him or her to record the numbers correctly in each column. A set of numbers might include 121, 346, 44, 12, 97, 9,876, 3,748, 220, 9,342, and 7,754.

Place-Value Cards

Distribute the *Place-Value Cards* (placevaluecards.pdf) to a small group of three to four students, giving each player four cards. The goal is to create the highest-value four-digit number. Students must try to get at least one card of each place value so that a four-digit number can be made. Taking turns, each player can ask for up to three new cards. From the (up to seven) cards in each player's possession, they should create the largest four-digit number possible.

Add-Visor

Write the words *thousands*, *hundreds*, *tens,* and *ones* on each of four sun visors. Put four students in a line, wearing the visors, and give each student a number (1–9) to hold. Have another student organize the students into the correct order, and then read the number. Have fun switching visors and numbers.

Show Me the Money

Using play money to represent thousand-, hundred-, ten-, and one-dollar bills, verbally give the student a three- or four-digit dollar value, such as $5,772. Allow the student to write it if needed. Then, ask him or her to take the correct bills to show that value.

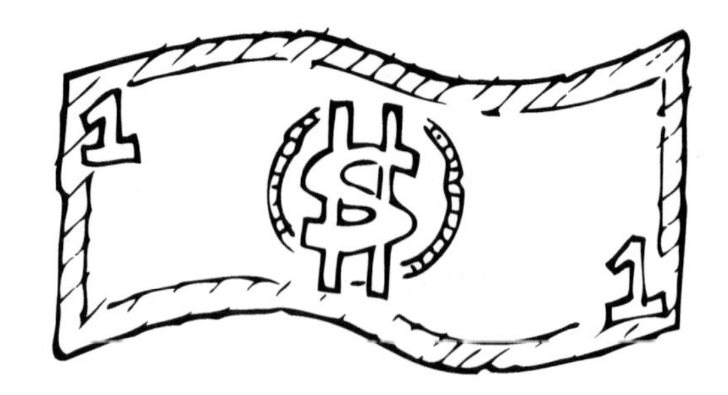

Dollars, Dimes, and Pennies *(Challenge Activity)*

Give the student a three-digit money value, such as $4.92. Using play dollar bills, dimes, and pennies, ask him or her to take the correct dollars and coins to represent that value.

Variation: Distribute *Number Dollars, Dimes and Pennies Array* (page 163) to the student. Give the student a three-digit money value, such as $6.25. Have him or her draw pictures or write numerals in the *Dollars*, *Dimes*, and *Pennies* columns to represent the amount.

Skills and Activities for Proficiency in Grade 3

By third grade, if students have mastered the essential math skills sequence for grades Pre-K–2, most students will have developed a solid understanding of basic mathematics skills. These students will have a belief that they are capable of thinking and solving problems with numbers. It is time to use number sense and basic numeracy skills to learn higher-order applications of mathematics. Use rich math vocabulary while engaging the student in these activities (see *Math Vocabulary,* page 166).

Many of the essential third-grade math skills will be practiced with paper-and-pencil tasks, in combination with a variety of projects and activities. Take the time to build deep understanding. Teach to mastery. Allow the student to apply each skill to a variety of contexts. Build that love of math that will last for life.

Recommended Materials

- Number cubes
- Cardboard
- Recycled materials (e.g., wood, plastic, cardboard)
- Fabric
- Bottle of water
- Balance/scale
- Beanbags
- Chalk or tape
- Tape measure
- Sticky notes
- Meter stick
- Yardstick
- Butcher paper
- Thermometer
- Play money
- Various objects for a play store
- Playing cards
- Abacus
- Counters
- Base-ten units and rods
- Map
- Spinner, number cubes, or tiles
- Clothesline
- Metal washers
- Clothespins
- Chocolate chips
- Paper plates
- Egg carton or muffin pan
- Small objects (e.g., pennies, stones, beads, buttons, marbles, plastic chips
- Vinyl non-skid spots
- Flash cards
- Graph paper
- Checkerboard
- Index cards
- Objects from daily life (e.g., popcorn, jello, clay, sand)

Skills and Activities for Proficiency in Grade 3 *(cont.)*

Essential Math Skills

Skill 20: Reads and writes numbers to 10,000 in words and numerals

Skill 21: Uses common units of measurement: length, weight, time, money, and temperature

Skill 22: Adds/subtracts three-digit numbers on paper with regrouping

Skill 23: Rounds numbers to the nearest 10

Skill 24: Rounds numbers to the nearest 100

Skill 25: Adds/subtracts two-digit numbers mentally

Skill 26: Counts by 5, 6, 7, 8, 9, and 10, using manipulatives

Skill 27: Uses arrays to visually represent multiplication

Skill 28: Recognizes basic fractions

Skill 29: Solves written and oral story problems using the correct operation *(addition, subtraction, and grouping)*

Additional Resources

- *Place-Value Chart* (page 149; placevaluechart.pdf)
- *Greatest Sum Cards* (greatestsum.pdf)
- *Greatest Sum Board* (sumboard.pdf)
- *Spinner* (spinner.pdf)
- *Rounds Up or Down Hundred Chart* (page 153; roundsup.pdf)
- *Round to the Nearest Ten* (page 154; nearestten.pdf)
- *Round to the Nearest Hundred* (page 155; nearesthundred.pdf)
- *Mental Math Bingo* (page 156; mentalmath.pdf)
- *Fill in the Fraction* (page 157; fillinfraction.pdf)
- *Draw Fractions* (page 158; drawfractions.pdf)
- *Matching Fractional Values* (page 159; matching.pdf)
- *What Fraction Is Shaded?* (page 160; whatfraction.pdf)
- *What Fraction Is Shaded? 2* (page 161; whatfraction2.pdf)
- *Problem-Solving Bingo* (page 164; problemsolvingbingo.pdf)
- *Problem-Solving Bingo Story Problems* (bingoproblems.pdf)
- *Dollars, Dimes, and Pennies Array* (page 162; dollarsarray.pdf)
- *Number Dollars, Dimes, and Pennies Array* (page 163; numbersarray.pdf)
- *"Make 1" Cards* (make1.pdf)
- *Skill 29 Story Problems* (skill29problems.pdf)
- *Here's What I Think* (page 148; whatthink.pdf) *(optional)*

Grade 3 Skills

Skill 20

Reads and writes numbers to 10,000 in words and numerals

Skill Progression

Proficiency for this skill is demonstrated by the student's ability to read and write numbers to 10,000 in words and numerals with 100 percent accuracy on three or more days, using three or more activities.

Intervention	Developing	Proficient
Counts, reads, and/or writes numbers to 100	Reads and/or writes numbers to 1,000	Reads and writes numbers to 10,000

Activities for Skill Development and Assessment

By a Roll of the Number Cube

Give three number cubes to the student. Each number cube should be rolled once and then placed in the hundreds, tens, or ones place on the *Place-Value Chart* (page 149). With the number cubes in the correct order, have the student write out the value in words on the chart. As the student progresses, increase the difficulty of the task by using four or five number cubes to represent thousands and ten thousands.

Variation: Give three number cubes to a group of three students. Ask them to work together, and assign each a role as the hundreds, tens, or ones representative. Each student will roll the number cube and write the numeral in the correct position on the *Place-Value Chart*. The students should take turns writing the value in words, such as *three hundred fifty-six*.

Writing Progressions

Give the student a writing progression to help him or her understand how to write complex numbers. The main ideas to remember include:

- Counting and naming starts over again after every comma.
- Numbers from twenty-one to ninety-nine often use a dash between the numbers.
- Do not use *and* to connect words when writing whole numbers.

Use a progression such as the one below to help the student deeply understand the relationship between number values.

3: Three
23: Twenty-three
523: Five hundred twenty-three
1,523: One thousand, five hundred twenty-three
7,523: Seven thousand, five hundred twenty-three
37,523: Thirty-seven thousand, five hundred twenty-three

Break it Down

Help the student understand the fundamental structure of numbers by having him or her complete tables such as the one below.

Number	Number of Ones	Number of Tens	Number of Hundreds
300	300		
900		90	
600			6
500			

More or Less?

Give the student a number up to 10,000. Then, have him or her complete the tasks below. Repeat the activity with different numbers and similar patterns.

Begin with a number, such as 315.

Make it two tens more ____________________

Make it two hundreds more ____________________

Make it two thousands more ____________________

Make it two tens less ____________________

Make it two hundreds more ____________________

Make it two ten thousands more ____________________

Make it two hundreds less ____________________

Three Ways

Have the student practice writing numbers in three different ways: standard form, expanded form, and word form.

Standard Form: 3,128

Expanded Form: 3000 + 100 + 20 + 8

Word Form: Three thousand, one hundred twenty-eight

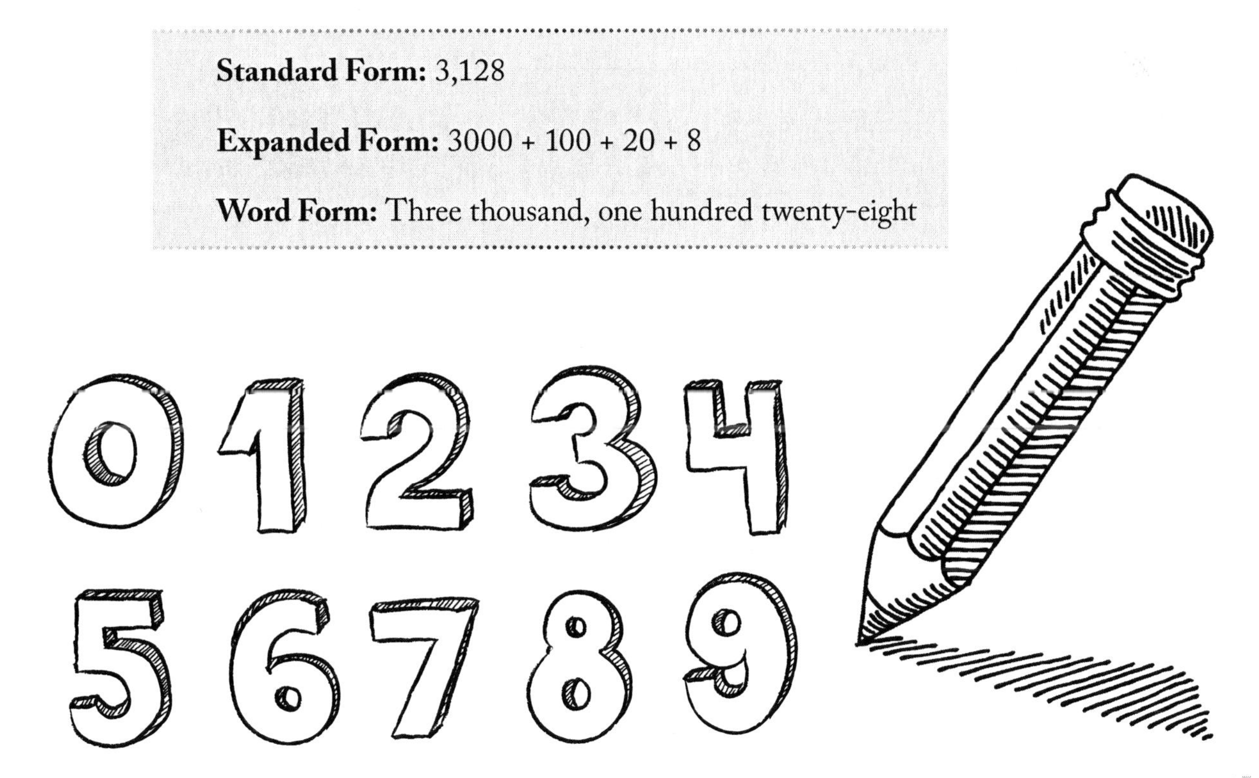

Grade 3 Skills

Skill 21

Uses common units of measurement: length, weight, time, money, and temperature

Skill Progression

Proficiency for this skill is demonstrated by the student's ability to accurately measure length, weight, time, money, and temperature on three or more days, using three or more activities. Specific minimal standards for proficiency include:

- Accurately measure the length of an object in inches (to 12 inches)
- Accurately measure the length of an object in feet (to 30 feet)
- Accurately measure the length of an object in yards (to 30 yards)
- Accurately measure the length of an object in centimeters (to 100 centimeters)
- Accurately measure the length of an object in meters (to 30 meters)
- Accurately measure weight in pounds (to 100 pounds)
- Accurately measure weight in ounces (to 16 ounces)
- Accurately measure weight in kilograms (to 100 kilograms)
- Accurately measure liquid in cups, pints, quarts, and liters (to 4 of each)
- Accurately tell time, using a digital or analog watch or clock
- Accurately time an event in minutes and seconds (to 30 minutes)
- Identify the value of money, using coins and bills (to $10)
- Make change for a purchase (to $1)
- Accurately identify the temperature of air or water, using both centigrade and Fahrenheit measures
- Accurately measure changes in temperature

Intervention	Developing	Proficient
Does not yet use units of measurement accurately	Able to use some units of measurement with accuracy	Consistently able to use all common units of measurement accurately

Activities for Skill Development and Assessment

Build It

Using cardboard or other available materials, have the student or a small team of students design and build a rocket ship, a castle, a fort, a race car, or another object of interest. Ask the student(s) to design it before beginning to build, emphasizing the use of measurement to plan ahead for materials.

Variation: Build a detailed structure with interlocking blocks or logs, and then measure every dimension of the finished structure.

Make a Measuring Stick

Using recycled materials such as wood, plastic, or cardboard, plan and build a measuring stick (e.g., ruler, yardstick, meter stick). Use this measuring stick to measure common objects to the nearest yard, foot, and inch, or meter and centimeter.

Potholders or Placemats

Design and create a potholder or placemat with geometric patterns, using colorful fabric. Measure the length of the sides of polygons to create balanced geometric designs.

Same or Different?

Cut pieces of cardboard into two different lengths. Ask the student to find and record objects in the classroom that are the same length as either of the cardboard pieces.

Using a small bottle of water containing a specific number of liquid ounces, ask students to predict which of several cups can hold the exact amount of liquid that is in the bottle. Experiment to see who is correct.

Using a simple balance or scale, find objects or groups of objects that have the same weight.

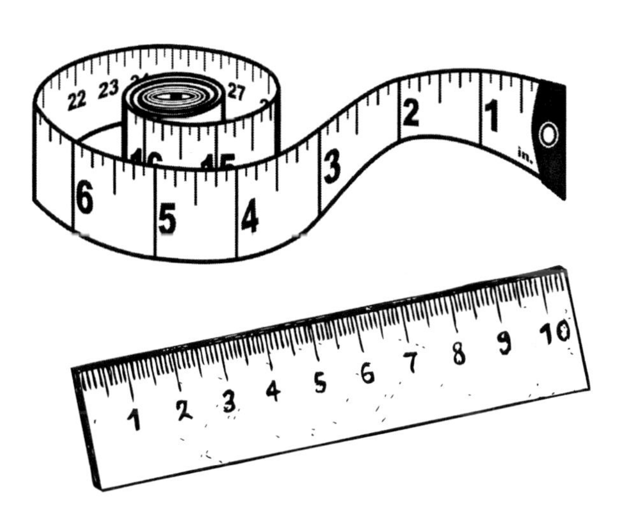

Beanbag Toss

Use chalk or tape to mark a starting line on the ground. Have students take turns tossing a beanbag. Measure and record the length of each student's toss in standard and/or metric units.

Measure Nature

With a flexible tape measure, record measurements of objects found in nature. For example, the student could measure the diameter of tree trunks, the length of shadows at different times of day, or the length of bugs and worms. Let measurement come to life. Expose the student to both metric and American standard units of measure.

Measure Sports

With a flexible tape measure and a scale, measure the dimensions and weight of every piece of sports equipment you can find.

Cook It

Use standard or metric measurements to help plan and cook a favorite dish. Use cups, pints, and quarts, or liters, for liquid ingredients; use ounces or grams for dry ingredients. Follow a recipe and cook something wonderful.

Eat It

Before diving into a snack, ask the student to weigh each portion, using a kitchen scale. If it is a snack that comes in a container, measure the dimensions. Record and eat!

Time It

How long will it take to walk across the classroom, go to the office, or circle the building? Plan destinations, and then have the student measure the time and the distance.

Cannot List

Make a list of everything you cannot measure using a measure of length, time, weight, temperature, or monetary value. Discuss the list.

Sticky-Note Measurement

Have the student calculate the area of a surface in square inches by first covering it in sticky notes and then using the sticky notes' dimensions of 3 × 3 inches.

Equivalent Measures

Using a meter stick and a yardstick, measure an object, using each tool, and record the different values. Discuss how units of measure are developed.

Silhouette Measurements

Using butcher paper (or chalk outside), outline the body of a student. Now measure the length and width of each body part. Compare results, using different students' silhouettes.

Measuring Furry Friends

Have the student bring in a stuffed animal from home. Measure every possible dimension of the toy. Graph the results, showing height, width, weight, length of ears, length of tail, etc.

How Much Time?

Make a list of time-related questions with the student, and allow him or her to choose three to five that he or she would like to answer. Questions might include *What time does your family have dinner? How long do you stay at the table? How long does it take to clean the dishes? How long does it take to do a load of laundry? How long does it take to get to school? What time does your favorite TV show come on? How long does it last? How much time during the show is used for commercials? What time does the sun go down?*

Hot and Cold

For a week, keep an hourly log of temperatures in the classroom and outside. Measure the temperature of various materials in the classroom, such as water in a jar, the windowsill, or inside a desk. Look for variations within the classroom.

Show Me the Money

Identify a monetary value (such as 24 cents) and ask the student to show you the coins that represent that value. Always design activities with just enough challenge to be fun, but not so much as to cause the student to become frustrated.

Count the Money

Using real or play coins, identify a value and have the student show that value with coins. Be sure to make this activity successful and fun for the student. Gradually teach the student to use the fewest possible coins.

Shopping

Establish a classroom store with priced objects. Ask the student to pick 1–3 objects and come to the register. Using play money, allow the student cashier to ring up the sale, take money from the buyer, and give correct change.

Budget Planning

Plan a field trip. Identify options for each planning decision (such as how much to spend on lunch and transportation), and allow the student to help plan the trip.

Plan an imaginary shopping trip. Give the student a limited budget and have him or her plan purchases.

Grade 3 Skills

Skill 22

Adds/subtracts three-digit numbers on paper with regrouping

Skill Progression

Proficiency for this skill is demonstrated by the student's ability to solve three-digit addition and subtraction problems on paper with 100 percent accuracy on three or more days, using three or more activities.

Intervention	Developing	Proficient
Unable to add or subtract three-digit numbers with regrouping	Able to accurately add and subtract three-digit numbers with guidance or use of manipulatives	Able to accurately add and subtract three-digit numbers on paper

Activities for Skill Development and Assessment

Three-Card Sums and Differences

Remove the tens and face cards from a standard deck of cards. Individually, in pairs, or in a small group, have students draw three cards to make a three-digit number. Students then draw a second set of three cards and add the two numbers to find the sum.

Variation: For subtraction, have students draw three cards to make a three-digit number and then draw a second set of three cards. Allow students to decide which of the two numbers is smaller and subtract that number from the greater number.

Greatest Sum

Prepare the *Greatest Sum Cards* (greatestsum.pdf) and distribute the *Greatest Sum Board* (sumboard.pdf) to students. Place the cards facedown between two players. Players take turns choosing a card until the six spots on their boards are filled and they have two three-digit numbers. The players should carefully decide where to place each card because they cannot move the cards once placed. After each player has six cards, each player should find the sum of their numbers. The player with the greatest sum wins.

Variation 1: Players can play "Smallest Sum," in which the object is to be the player with the smallest sum.

Variation 2: Players can play "Greatest Difference," in which the players will subtract the two numbers and the object is to be the player with the greatest difference.

Variation 3: Players can play "Smallest Difference," in which the object is to be the player with the smallest difference.

Spinner Addition and Subtraction

Using a spinner (spinner.pdf), paper clip, and a pencil to hold the paper clip in the center, have the student spin three times to make a three-digit number. Then, the student should spin a second set of three numbers and add the two numbers to find the sum.

Variation 1: Use the spinner to practice subtraction. Have the student spin three times to make a three-digit number and then spin a second set of three numbers. Allow the student to decide which of the two numbers is greater and then subtract the smaller number from it.

Variation 2: Use a number cube in place of the spinner for a similar activity.

Baseball Batting Averages

Engage a student who is interested in baseball by having him or her track the batting average of a favorite player through the course of the season. Check batting averages at the end of each week, and use subtraction to see the change in a player's performance.

Variation: Have the student choose a favorite team and determine which player has the best batting average and by how much.

Money, Decimals, and Computation

Money problems provide a great opportunity to connect meaning to addition and subtraction. Teach the student to add and subtract dollars, dimes, and pennies. Show him or her the value relationship between hundreds, tens, and ones.

Variation 1: Provide story problems using dollars, dimes, and pennies.

Variation 2: Using the *Dollars, Dimes, and Pennies Array* (page 162), ask the student to draw the bill or coin for each step in a money problem.

Dollars	Dimes	Pennies	Find the Sum
			\$3.21
			+ \$1.21
			\$4.42 total

Estimation Fun

Ask the student to look at a three-digit addition or subtraction problem, quickly estimate the answer, and then carefully solve the problem on paper. Design instruction so that it is just challenging enough to be fun for the student as he or she hones estimation skills.

Story Time

Read the student a story problem. Include one piece of unnecessary information. Have the student figure out which numbers are relevant to the solution, choose the correct operation, and solve the problem. For example: *James and Julianne travelled 205 miles to their grandmother's house for Thanksgiving. They stopped once along the way to pick up Great-Aunt Beatrice, who is 91 years old. They stayed overnight and then drove all the way home, leaving Aunt Beatrice with Grandma. How many miles did they travel in these two days?*

Grade 3 Skills

Skill 23

Rounds numbers to the nearest ten

Skill Progression

Proficiency for this skill is demonstrated by the student's ability to round numbers (up to four digits) to the nearest 10 with 100 percent accuracy on three or more days, using three or more different activities.

Intervention	Developing	Proficient
Unable to round numbers to the nearest ten	Inconsistently rounds numbers to the nearest ten	Consistently rounds numbers to the nearest ten

Activities for Skill Development and Assessment

Rounding Chart

Identify a number on the *Rounds Up or Down Hundred Chart* (page 153). Ask the student to place a counter on that number and then round it up or down to the nearest 10. As the student gains understanding with the concept of rounding, ask him or her to touch the number with his or her finger before rounding up or down to the nearest 10.

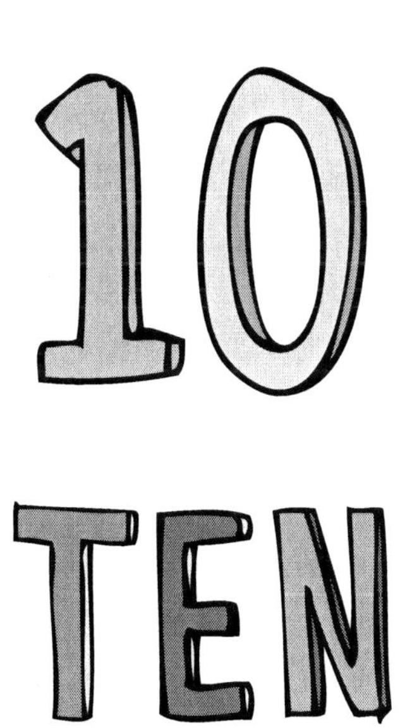

Abacus Rounded

Using an abacus, ask the student to show you a number (to 100) and then round that number to the nearest 10. Have him or her show the original value on the abacus and then show the rounded value. Ask the student to justify why he or she has rounded to the specified 10 and use the abacus to support the response.

Around and About

Using the *Rounds Up or Down Hundred Chart* (page 153) and counters, ask the student to identify an increment of 10, such as 40, and then place counters on every number that will round up or round down to that number.

Rounding Money

Using real or play coins, ask the student to show you a value up to 100 (such as 67 cents). Then, ask him or her to round up or down to the nearest ten. Have the student explain why he or she chose that answer.

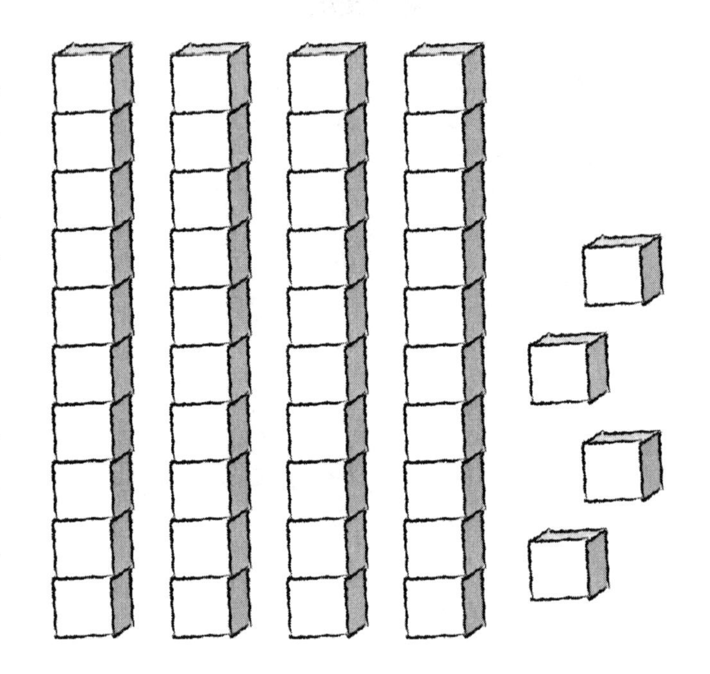

Units and Tens

Using base-ten units and rods, ask the student to show a value up to 100 (such as 44). Then, ask him or her to round up or down to the nearest ten.

Round to the Nearest Ten Chart

Distribute the *Round to the Nearest Ten* activity sheet (page 154). Fill in number values in the left column. Ask the student to write the value of thousands, hundreds, tens, and ones in the appropriate columns and then round to the nearest ten.

What Numbers Can Round to This Ten?

Identify a certain number of tens (e.g., 9 tens). Then, ask the student to write down every number that rounds up or down to that number (e.g., 91, 92, 93, 94, 89, 88, 87, 86, 85). Use larger numbers as appropriate (e.g., 48 tens). Ask the student to write every number that rounds up or down to that number (e.g., 481, 482, 483, 484, 479, 478, 477, 476, 475).

Grade 3 Skills

Skill 24

Rounds numbers to the nearest hundred

Skill Progression

Proficiency for this skill is demonstrated by the student's ability to round numbers (up to five digits) to the nearest 100 with 100 percent accuracy on three or more days.

Intervention	Developing	Proficient
Unable to round numbers to the nearest hundred	Inconsistently rounds numbers to the nearest hundred	Consistently rounds numbers to the nearest hundred

Activities for Skill Development and Assessment

Round to the Nearest Hundred Chart

The student who completely understands rounding to the nearest ten will be ready to understand rounding to the nearest hundred. Using the *Round to the Nearest Hundred* activity sheet (page 155), carefully teach the student the procedure for rounding to the nearest hundred.

Map it Out

Ask the student to name several cities that he or she would like to visit. Using a digital or paper map, figure the distance and round to the nearest hundred miles.

Car Costs

Have the student determine the price of several of his or her favorite cars. Have the student round this price to the nearest hundred.

Money Matters

Using play money, ask the student to show you a value up to $999. Then, ask him or her to round up or down to the nearest hundred and explain why he or she chose that answer.

Rounding Practice Fun

Using a spinner, number cubes, or tiles with numerals, have the student pick three number values and place them in sequence. Ask the student to round to the nearest hundred and explain why he or she chose that answer.

Variation: Have the student pick 4–5 number values in sequence and then round to the nearest hundred.

Grade 3 Skills

Skill 25

Adds/subtracts two-digit numbers mentally

Skill Progression

Proficiency for this skill is demonstrated by the student's ability to add and subtract two two-digit numbers (with sums up to 100) mentally with 100 percent accuracy on three or more days, using three or more different activities.

Intervention	Developing	Proficient
Unable to mentally add or subtract two-digit numbers	Sometimes able to demonstrate ability to mentally add and/or subtract two-digit numbers	Consistently demonstrates ability to mentally add and subtract two-digit numbers

Activities for Skill Development and Assessment

Add Along

Give the student an easy number to add on to, such as 20. Then, give the student a two-digit number to add to it, and ask him or her to write the sum on a sheet of paper. After the student has experienced some success with this activity, gradually increase the difficulty. Challenge the student just enough, but don't push him or her to frustration. Aim for success 90 to 95 percent of the time.

Easier: 23 + 20, 44 + 10, 88 – 20

Moderate: 15 + 16, 81 – 11, 22 + 22

Challenge: 56 + 25, 78 – 19, 99 – 67

Variation: After the student has mentally determined a sum, have him or her show the solution on the abacus.

Doubles

Give the student a two-digit number, and ask him or her to double it. Start with numbers that do not require regrouping (e.g., 11, 12, 13, 14), and work up to harder numbers (e.g., 26, 42, 49). Add levels of difficulty when the student is ready for more of a challenge:

Second level: "I'll say a number, you double it. Then, add [a number]."

Third level: "I'll say a number, you double it. Then, subtract [a number]."

Fourth level: "I'll say a number. You add [a number], and then double the sum."

Mental Math Warm-Up

Begin a whole-group or a small-group lesson with this mental-math warm-up. Adapt the level of difficulty to the readiness level of the class or group. Begin with a three- or four-step math sequence for most groups. For example, "Start with 6. Add 27. Subtract 9. Raise your hand when you have the solution." Deliver the steps slowly at first, developing listening skills as well as mental-math skills. Over the course of the year, increase the speed and complexity of the problem.

Story Problems

Write two- or three-step story problems that are just challenging enough for the student. Ask him or her to solve these addition and subtraction problems without the use of a pencil and paper. Deliver the story problems orally. Ask the student to listen to a problem and then solve it mentally.

Basketball Sports Update

Deliver an update on a fictitious basketball game at the end of each imaginary quarter. Describe the action from each quarter and tell the student how many points were scored by each team, but do not give the student the running score. Allow the student to keep score in his or her head as the information is delivered.

Taking Measure

Move around the room with a meter stick or a yardstick, stopping to measure different objects. Call out the number of centimeters or inches an object measures, and allow the student to add the numbers in his or her head.

Checkout Challenge

Pretend that you are checking out and paying for purchases at a department store. Using whole-dollar amounts (no cents), identify the cost of each item, and have the student figure out the total bill.

Mental Math Bingo

Distribute the *Mental Math Bingo* activity sheet (page 156). Read the directions to the student, and then give him or her a starting number between 25 and 100. The student should find the quickest and easiest solutions that will allow him or her to answer five questions in a row. The correct answers can be vertical, horizontal, or diagonal. All computation must be done mentally.

Grade 3 Skills

Skill 26

Counts by 5, 6, 7, 8, 9, and 10, using manipulatives

Skill Progression

Proficiency for this skill is demonstrated by the student's ability to quickly skip-count by 5, 6, 7, 8, 9, and 10 with 100 percent accuracy on three or more days, using three or more different activities.

Intervention	Developing	Proficient
Unable to skip count using manipulatives	Skip counts by 5 and 10, but struggles with 6, 7, 8, or 9, using manipulatives	Consistently counts by 5, 6, 7, 8, 9, and 10, using manipulatives

Activities for Skill Development and Assessment

Absolutely Abacus: Skip Counting 1

Using the abacus, ask the student to count by a certain number (5, 6, 7, 8, 9, or 10) to a multiple of 10 (e.g., count by 8 to 80). First, he or she moves 8 beads and then says *eight*. Then, the student moves another set of 8 and now says *sixteen*. Every time he or she says a number, that number should be clearly shown on the abacus, using every bead in the top row before moving down to the next row. Practice several number sequences per day, and have fun with the numbers. Give the student many days of practice to build understanding of these important number sequences.

Clothesline Counting

Hang a clothesline on which 100 metal washers have been strung. Ask the student to slide groups of 5, 6, 7, 8, 9, or 10 washers down the clothesline as he or she counts by 5, 6, 7, 8, 9, or 10. The student can use clothespins to separate the groups of washers.

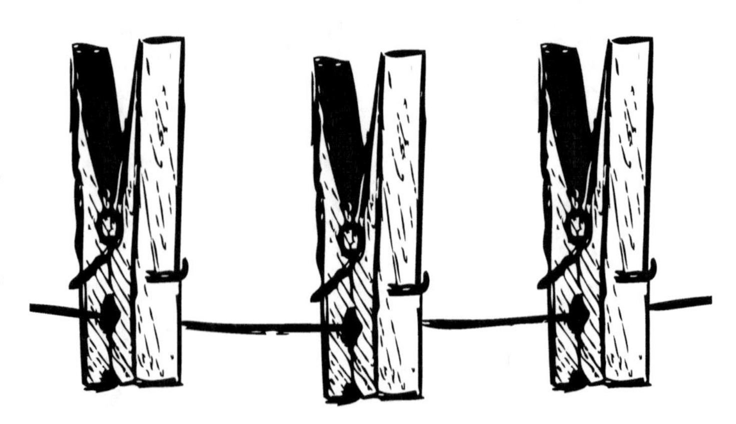

Chocolate Counts

Pour a pile of chocolate chips or chocolate candies onto a paper plate. Ask the student to count the group by a certain multiple (e.g., 5, 6, 7, 8, 9, 10) and then set aside (or eat) any leftovers. Then, pick a different number to use for skip counting the chocolate, and continue to set aside or eat the leftovers until the pile is gone.

Skip-Count McStuff

Using egg cartons or muffin pans, count out a number of objects (e.g., buttons, beads, marbles, chips, stones) for each space. For example, you might place 6 objects in each section of the muffin pan and count by 6 to 60.

Get in Line

Place a line of 10 vinyl non-skid spots on the floor. Ask the student to walk on the spots and count by a certain number with each step.

Variation: With a small group of students, give each student a number card (e.g., 5, 6, 7, 8, 9, or 10), and then ask each student to walk the path while skip counting by that number. When complete, ask students to swap cards and do it again.

Plates

Create sets of 10 small paper plates on which 5, 6, 7, 8, or 9 symbols (e.g., circles, stars, hearts, diamonds, squares) have been drawn. Give one set to the student, and have him or her count by that number, passing you one plate at a time. Try the same activity with the numbers 5, 6, 7, 8, 9, and 10.

Absolutely Abacus: Up and Down Skip Counting

Once the student is skilled at using the abacus to count up by 6, 7, 8, 9, and 10, introduce counting backward to zero, showing the correct set of beads as you say each number. Have the student count to different end points. For example, count by 8 to 32; count by 5 to 30; or count backward by 9 from 63 to 18.

Absolutely Abacus: Show Me Groups

Ask the student to show a certain number of groups of a specific number, such as three groups of five. He or she will then move groups of five on the abacus, using the entire top row of beads before going on to the next row. Once the student has correctly moved three groups of five, ask him or her to identify the total by asking, "How much do you have?" Practice this new skill, using problems that allow at least 90 percent success. Repeat this grouping activity at least 10 times in a session. Gradually increase the difficulty of this task as the student gains skill, using all combinations of numbers up to 10.

This skill establishes a basis for understanding multiplication. By mastering these exercises, your student will never struggle to memorize multiplication facts without understanding the meaning of multiplication. Instead, he or she will be a student who truly understands grouping. Do not instruct the student to multiply 3 × 5; instead, purposefully use more concrete language such as "Show me three groups of five."

Test the Teacher

As before, let the student make up challenging problems for the teacher. If you (the teacher) answer five questions correctly, ask the student to write a note indicating that you performed at a high level of skill.

Activities to Extend Deep Understanding

Absolutely Abacus: In a Group

Note: *This exercise is a concrete representation of division, but there is no need to discuss this with the student. Use the concrete language below to ensure deep understanding of the concept.*

Ask the student to show a certain number on the abacus, such as 35. Once done, ask him or her, "How many groups of 5 are in 35?" The student can separate the groups slightly or use any inventive technique to help him or her identify the number of groups. Celebrate success. Repeat similar exercises at least 10 times in a session. Use easy problems at first (e.g., those with factors of 2, 5, or 10). Gradually increase the difficulty of the problems. Give the student several weeks to practice this activity before expecting competence.

Absolutely Abacus: Fractional Groups *(Halves)*

Ask the student to show you a certain number on the abacus. (Begin with an even number up to 20.) Start with the easily understood concept of halves. Ask, "What's half of this number?" Move toward using larger numbers up to 100. This activity begins to teach the student that the language of fractions is just another way of looking at groups. Practice at least 10 repetitions in a session. Occasionally offer story problems to be sure that the student understands the uses of fractional language.

Absolutely Abacus: Fractional Groups 2 *(Quarters, Thirds, and Fifths)*

Ask the student to show you a number that can be divided evenly into fractional parts. Ask the student to show one-fourth, one-third, or one-fifth of the number. For example:

Show me 12. Now, divide 12 into quarters. How many beads are in each quarter? (3)

Show me 10. Now, divide 10 into fifths. How many beads are in each fifth? (2)

Use small numbers until the student completely understands this interesting concept. Then, enjoy challenging him or her with more complex problems using numbers up to 100. Occasionally use story problems to make real-life connections.

Absolutely Abacus: Grouping on Paper

Ask the student to solve a simple grouping problem on the abacus and then on paper. Sample problems are listed below. Allow plenty of time to explore deep understanding of the meaning of both multiplying and dividing so that every student will completely understand these processes.

"Show me [a number] groups of [another number]. Great—now how many do you have altogether?" Once the student has shown the correct answer on the abacus, ask him or her to record the problem on paper. The student might write *3 groups of 5 is 15.* Use words to describe the problem until you are sure that the student understands how to use the multiplication symbol, and then allow him or her to write $3 \times 5 = 15$.

"Show me [a number]. Correct! Now, how many groups of that number are in [another number]?" Once the student has shown the correct answer on the abacus, ask him or her to record the problem on paper. The student might write *35 is equal to 5 groups of 7.* Use words to describe the problem until you are sure that the student understands how to use the symbols. Then, allow him or her to write $35 = 5 \times 7$.

"Show me [a number]. Correct. Now, divide that number into [another number] groups. How many are there in each group?" Once the student has shown the correct answer on the abacus, ask him or her to record the problem on paper. The student might write *35 divided into 5 groups makes 7 in each group.* Use words to describe the problem until you are sure that the student understands how to use the division symbol, and then allow him or her to write $35 \div 5 = 7$.

"Show me [a number]. Correct. Now, divide that number into groups of [another number]. How many groups will there be?" Once the student has shown the correct answer on the abacus, ask him or her to record the problem on paper. The student might write *22 divided into groups of 2 gives you 11 groups.* Use words to describe the problem until you are sure that the student understands how to use the division symbol, and then allow him or her to write $22 \div 2 = 11$.

Skip-Counting Catch

When the student has mastered skip counting by any base number, consider a catching game to improve speed and automaticity. Using a beanbag, play underhand catch with the student. Each time the student catches the beanbag, he or she must say the next number in the progression.

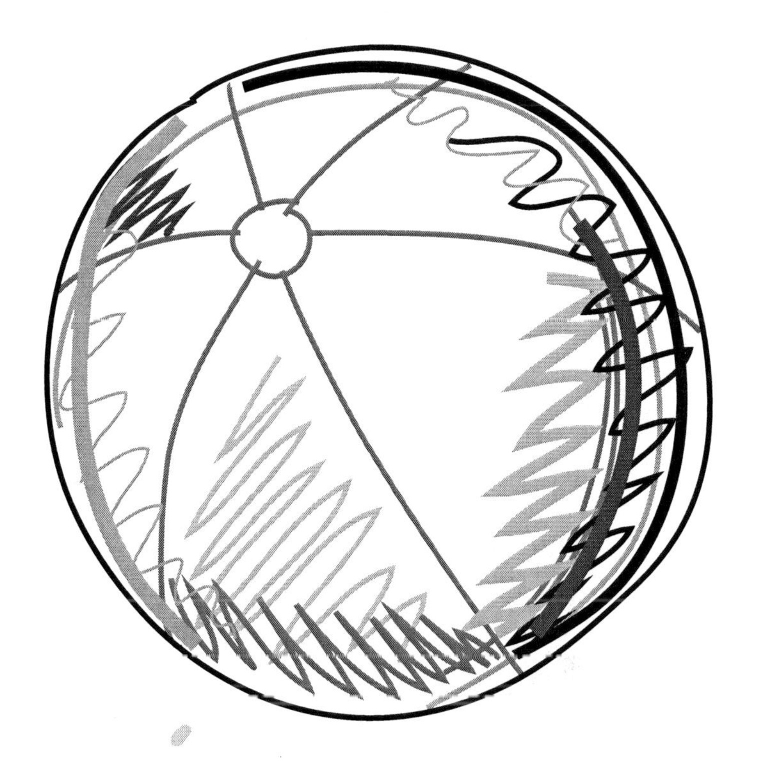

Skip-Counting Group Catch

Gather a small group of 3–5 students in a circle. Using a beanbag or a large, bouncy ball, play catch with the students. Each time a student catches the ball, he or she must say the next number in the progression (e.g., *6, 12, 18, 24, 30, 36, 42, 48, 54, 60, 54, 48, 42, 36, 30, 24, 18, 12, 6, 0*).

Flashcard Skip Counting

Note: *It is important to help the student develop a deep understanding of numbers, sequences, and groups before asking him or her to memorize multiplication facts. Students should not memorize facts without deeply understanding the value of the numbers involved.*

Show the student a multiplication fact flashcard, such as 7 × 5, and in response have the student count by 7 or 5 to 35. Use flashcards for any multiplication fact the student has come to deeply understand. Once the student can complete this activity with ease, typical activities with flashcards can be used without hindering conceptual understanding.

Grade 3 Skills

Skill 27

Uses arrays to visually represent multiplication

Skill Progression

The array is a very powerful model for supporting the development of students' thinking around both multiplication and division. Teach the terms *row* and *column*. Also, ask the student to think of the multiplication symbol as having the meaning "rows of." This way, when the student thinks of 3 × 5, he or she will think of 3 rows of 5.

Proficiency for this skill is demonstrated by the student's ability to draw or build multiplication arrays with 100 percent accuracy on three or more days, using three or more different activities.

Intervention	Developing	Proficient
Uses arrays to visually represent multiplication problems with a factor of 2	Uses arrays to visually represent multiplication problems with factors through 6	Uses arrays to visually represent multiplication for problems with factors through 12

Activities for Skill Development and Assessment

Building Arrays

Have the student use blocks, bingo chips, blank scrabble tiles, buttons, pennies, or other physical objects to build arrays. Ask the student to build an array with a certain number of rows and columns. Practice building arrays many times to give the student the repetition needed to deeply understand this concept.

Advanced level: Have the student write an equation or a story problem to match an array.

Graph-Paper Arrays

Ask the student to draw an array on graph paper with a certain number of rows and columns. Then, ask the student to write an equation or a story problem to match the array.

Symbolic Arrays

Ask the student to draw an array of symbols with a certain number of rows and columns. The student might use symbols such as circles, squares, smiley faces, or stars. Then, ask the student to write an equation or a story problem to match the array.

Abacus Arrays

Using the abacus, ask the student to build an array with a certain number of rows and columns. Then, ask the student to write an equation or a story problem to match the array.

Roll the Number Cube

Ask the student to follow these steps:

Step 1: Roll the number cube twice. The first number you roll tells how many rows to make in your array. The second number you roll tells how many columns to make.

Step 2: Draw the array. For example, if you rolled 3 first and then a 4, you would draw this array:

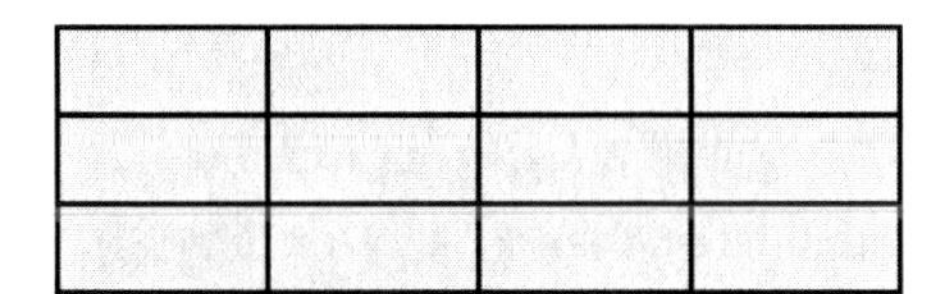

Step 3: Write an equation to match your array.

Array Picture Cards

Prepare a set of picture cards showing various arrays. For fun, make arrays using different symbols and pictures. Show the student one picture card. Ask him or her to identify the number of rows and columns. Once the student has correctly identified these values, ask him or her to do each of the following:

1. Write a multiplication equation to match your card.
2. Write a multiplication story problem to match your card.

Advanced level:

1. Write a division equation to match your card.
2. Write a division story problem to match your card.

Human Arrays

On the playground, draw a grid with chalk. Ask a group of students to quickly create an array with a certain number of rows and columns by placing themselves in the grid. Look for ways to have fun with arrays. Create teams to compete for speed in creating arrays, or use music.

Story Problem Arrays

Ask the student to solve written or oral multiplication story problems by first drawing out the appropriate array and then writing the equation.

Checkerboard Arrays

Ask the student to solve written or oral multiplication story problems by first showing the array on a checkerboard and then writing the equation.

Grade 3 Skills

Skill 28

Recognizes basic fractions

Skill Progression

Proficiency for this skill is demonstrated by the student's ability to identify, draw, and write basic fractions including $\frac{1}{2}$, $\frac{1}{3}$, and $\frac{1}{4}$ with 100 percent accuracy on three or more days, using three or more different activities.

Intervention	Developing	Proficient
Unable to identify $\frac{1}{2}$, $\frac{1}{3}$, and $\frac{1}{4}$ in picture form with automaticity	Able to identify basic fractions in pictures and draw basic fractions	Able to identify, draw, and write basic fractions

Activities for Skill Development and Assessment

Building Fractions

Use manipulatives such as interlocking cubes, bingo chips, blank letter tiles, buttons, fraction tiles, coins, play pizza, or other physical objects to teach the basic concept of fractional parts. Using small groups of objects, ask the student to show you the value of $\frac{1}{2}$, $\frac{1}{3}$, and $\frac{1}{4}$.

Fill in the Fraction

Complete the directions on the *Fill in the Fraction* activity sheet (page 157) with either $\frac{1}{4}$, $\frac{2}{4}$, $\frac{3}{4}$, or $\frac{1}{2}$, and ask the student to practice shading fractional parts. The activity sheet helps the student see that fractions can represent parts of one object or parts of groups.

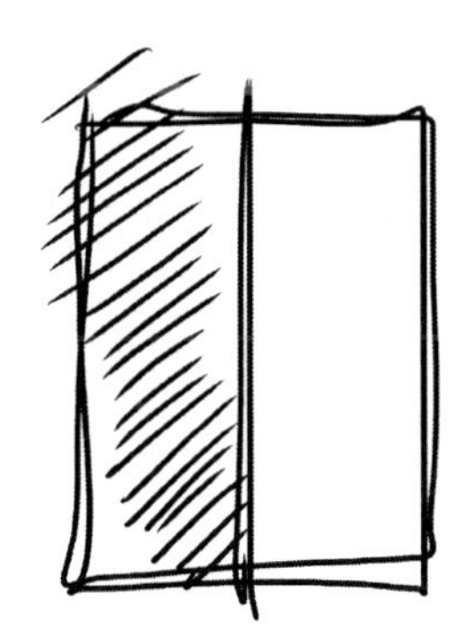

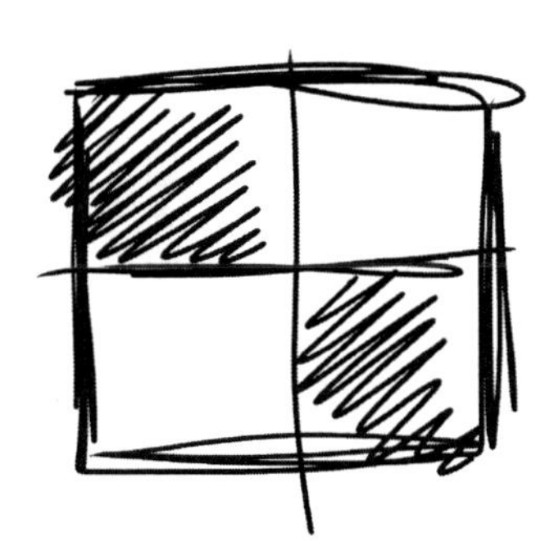

Fair Share

Give the student story problems in which he or she must decide how to give each character an equal share. For example:

Molly and Matt both want the last cookie. Show how to divide the cookie so that each gets an equal share.

Molly and Matt each want more grapes, but there are only six grapes left. Show how to divide the grapes so that each gets an equal share.

Ask the student to write story problems representing an "equal share" scenario.

Name That Fraction!

Draw visual representations of basic fractions on index cards. Use these for short speed drills by flashing a card and asking the student to name the fraction shown. **Caution:** Use speed drills only after the student has developed deep understanding of the mathematical concept. Understanding should precede speed.

Use fractions represented by different forms:

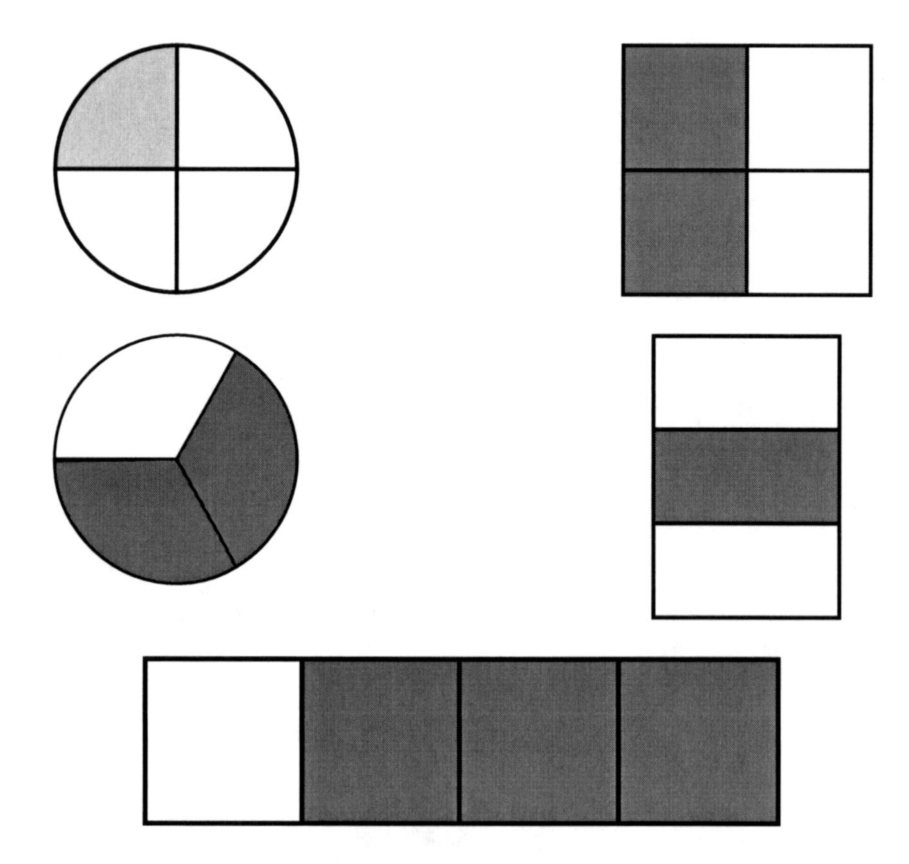

Estimating Fractions

Sometimes fractions are not measured in discreet measures, but rather by estimates. Using water, sand, clay, jello, and other interesting substances, have the student fill or divide something from daily life in a way that represents fractions. Make fractions real, and make them fun. For example: *Fill half the bottle with water; Drink half the orange juice; Give each player one-quarter of the clay; Divide the jello into three portions; Serve the popcorn to four people; Fill half the box with dirt.*

Pizza Fractions

Ask the student to draw his or her favorite pizza on a paper plate and then cut it into slices to show halves, thirds, or fourths.

Matching Fractional Values

Help the student understand the different representations of a fractional value by using the *Matching Fractional Values* activity sheet (page 159). The student will look for matching representations such as:

$\frac{1}{4}$ one-fourth

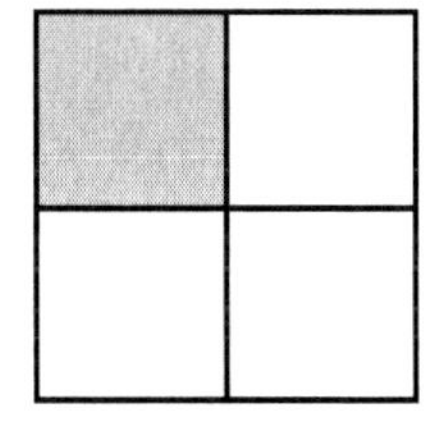

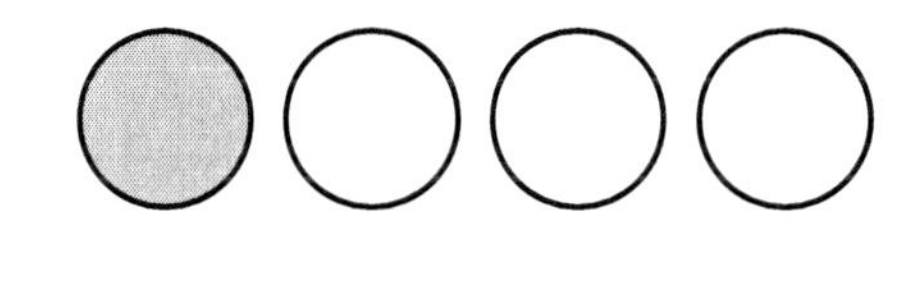

"Make 1" Card Game

Prepare the *"Make 1" Cards* (make1.pdf). The object of the game is to collect any combination of cards that add up to exactly one whole. With two to four players, follow these rules:

1. Give 3 cards to each player, dealing one card at a time to each player.
2. As soon as you have cards adding up to exactly one whole, show the cards and say "One." That player earns one point.
3. If no player has one whole, each player is dealt another card.
4. This pattern continues until someone wins the hand and gets one point.
5. Deal another hand to each player, and the game continues.
6. The first player with five points wins.

Activities to Extend Deep Understanding

Draw Fractions, Advanced

Distribute the *Draw Fractions* activity sheet (page 158). Ask the student to complete the activity sheet by drawing fractional values using circle, square, or rectangular wholes, or sets.

Design Fraction Flags

Have the student find, or design and draw, a national flag or a signal flag that uses simple fractions. Encourage the student to write sentences that describe the fractions (e.g., *One-third of the flag is green*).

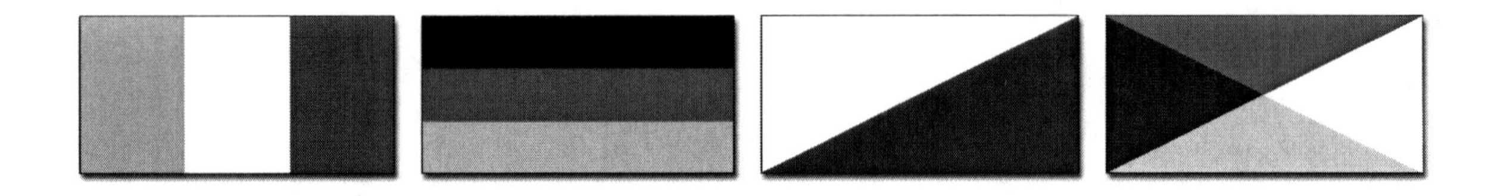

What Fraction Is Shaded?

Distribute the *What Fraction Is Shaded?* activity sheet (page 160) to the student. Ask the student to identify the fraction of each rectangle that is shaded.

Variation: Use *What Fraction Is Shaded? 2* (page 161) to create your own activity sheet for students. At first, use values that represent $\frac{1}{2}$, $\frac{1}{3}$, and $\frac{1}{4}$. Once this becomes easy, use more complex fractional values (e.g., $\frac{2}{4}$, $\frac{3}{9}$, $\frac{4}{12}$).

Grade 3 Skills

Skill 29

Solves written and oral story problems using the correct operation *(addition, subtraction, and grouping)*

Skill Progression

Proficiency with this skill is demonstrated by the student's ability to identify the correct operation and solve both written and oral story problems. These can include story problems that require two-digit addition, two-digit subtraction, or basic grouping. The student is able to develop a plan, use the correct operation, and solve 10 story problems with 100 percent accuracy on three or more days.

Intervention	Developing	Proficient
Able to solve written and/or oral story problems if given plan and correct operation	Able to solve written and oral story problems with guidance in developing plan	Consistently able to solve written and oral story problems by developing a plan, using correct operations, and evaluating the solution

Activities for Skill Development and Assessment

Draw It Out

Story problems can be simple: *Melinda had 6 apples. Then, she bought 4 more. How many apples does she have now?* Or they can be wonderfully complex: *Brad, Tyrel, and Alicia played marbles. The game went badly for Brad. When the game began, he had 22 marbles. At the end of the game Brad had only 4 marbles left, while Tyrel had three times as many. And Alicia had three times as many as Tyrel. While walking home, Alicia found 2 more marbles lying on the sidewalk. At the end of the day, how many marbles does each child have?*

Teach the student to illustrate story problems to ensure full understanding. Story problems can be illustrated by drawing out figures to represent the values in the problem.

Consider allowing the student to work with a partner on story problems, or give the student choices as to which story problem he or she would like to illustrate. Sample story problems are provided below.

The holiday tree in Glen Haven is 40 feet high. The man who puts the star on the top is 6 feet tall, but he can reach 1 foot above his head. If he stands 5 feet below the top of the ladder, how high must his ladder be for him to reach the top of the tree?

There are 3 shelves of books. Two of the shelves hold 25 books each. One of the shelves holds 37 books. How many books are on the shelves altogether?

There are 24 students in Mrs. Monroe's class. Half of the students are wearing running shoes. Five students are wearing sandals. Five students are wearing boots. How many running shoes are in the classroom?

Listen and Solve

Listening to a story problem helps students practice visualization and memory. Read story problems aloud to the student. Consider partnering students and allowing them to help each other.

Variation: Pair students. Have one student read a story problem silently, cover the print, and then explain the problem one time to his or her partner. The partner then follows the directions and solves the problem on paper.

Write Story Problems

A student who writes good story problems can solve story problems. Teach the student to write math story problems and to use these guidelines:

- Story problems should include irrelevant data. Include numbers or descriptions that are not needed for the solution to the problem.
- Story problems should use a variety of types of questions. Using one type of question gets boring and predictable.
- Story problems should not use the same operation (adding, subtracting, or grouping) repeatedly. Vary the type of problem so the student must think about how to solve it.
- Story problems should be interesting. Good writing matters.

The Story-Problem Center

Create a story-problem center in the classroom. Write story problems that use each of the essential math skills. Organize the problems by theme and difficulty. Have manipulatives available so the student can use these or visual mapping to understand the problem and find the solution. Consider asking the student to write one high-quality story problem as the "price of admission" to the center. Teach students how to collaborate and solve challenging problems as a team. Make going to the story-problem center a preferred activity.

Problem-Solving Bingo

Use the *Problem-Solving Bingo* activity sheet (page 164) to create bingo cards by writing the answers to 25 short story problems in each square. Use the *Problem-Solving Bingo Story Problems* (bingoproblems.pdf), if desired. Prepare several different versions of the bingo card by rearranging the answers on the cards.

Working with a small group of students who are at similar readiness levels, explain the rules of the game: The teacher will read one story problem at a time, and then allow time for students to solve the problem and find the answer on their bingo card. Students should place a marker on the correct answer. This will repeat until the first player to have five correct answers in a row vertically, horizontally, or diagonally has bingo. The winner will verify the answers.

Variation 1: Distribute the bingo cards and 25 story problems to each student. Have students read and solve the problems independently. The first student to find five correct answers in a row wins.

Variation 2: Distribute a bingo card and 25 story problems to the student. Allow him or her to choose and solve problems until he or she has bingo. Ask the student to verify and/or explain the answers.

Story Problem Practice

Distribute the *Skill 29 Story Problems* (skill29problems.pdf) to the student for additional practice with grade-level story problems.

The Pre-K to Grade 3

Essential Math Skills Individual Inventory

Directions: Use the *Skill Progression Rubrics* (Appendix A) to evaluate the student's performance on each skill.

Student: ______________________________ Date: ________________

Skill	Not Yet Assessed	Emerging/ Intervention	Developing	Proficient
Skill 1: Demonstrates one-to-one correspondence for numbers 1–10, with steps				
Skill 2: Demonstrates one-to-one correspondence for numbers 1–10, with manipulatives				
Skill 3: Adds on, using numbers 1–10, with steps				
Skill 4: Adds on, using numbers 1–10, with manipulatives				
Skill 5: Demonstrates counting to 100				
Skill 6: Has one-to-one correspondence for numbers 1–30				
Skill 7: Understands combinations (up to 10)				
Skill 8: Recognizes number groups without counting (2–10)				
Skill 9: Counts objects with accuracy to 100				
Skill 10: Replicates visual or movement patterns				
Skill 11: Understands concepts of adding on or taking away (to 30), with manipulatives				
Skill 12: Adds/subtracts single-digit numbers on paper				
Skill 13: Shows a group of objects by number (to 100)				
Skill 14: Quickly recognizes number groups (to 100)				
Skill 15: Adds/subtracts from a group of objects (to 100)				

The Pre-K to Grade 3
Essential Math Skills Individual Inventory *(cont.)*

Skill	Not Yet Assessed	Emerging/ Intervention	Developing	Proficient
Skill 16: Adds/subtracts two-digit numbers on paper				
Skill 17: Counts by 2, 3, 4, 5, and 10, using manipulatives				
Skill 18: Solves written and oral story problems using the correct operations *(addition and subtraction)*				
Skill 19: Understands/identifies place value to 1,000				
Skill 20: Reads and writes numbers to 10,000 in words and numerals				
Skill 21: Uses common units of measurement:				
Length				
Weight				
Time				
Money				
Temperature				
Skill 22: Adds/subtracts three-digit numbers on paper with regrouping				
Skill 23: Rounds numbers to the nearest ten				
Skill 24: Rounds numbers to the nearest hundred				
Skill 25: Adds/subtracts two-digit numbers mentally				
Skill 26: Counts by 5, 6, 7, 8, 9, and 10, using manipulatives				
Skill 27: Uses arrays to visually represent multiplication				
Skill 28: Recognizes basic fractions				
Skill 29: Solves written and oral story problems using the correct operation *(addition, subtraction, and grouping)*				

Pre-Kindergarten
Essential Math Skills Class Inventory

Directions: List student names in the first column. When a student reaches proficiency with a skill, record the date in the column under that skill. Update the inventory each week.

	Pre-Kindergarten Skills				Kindergarten Skills			
Student Name	**Skill 1:** Demonstrates one-to-one correspondence for numbers 1–10, with steps	**Skill 2:** Demonstrates one-to-one correspondence for numbers 1–10, with manipulatives	**Skill 3:** Adds on, using numbers 1–10, with steps	**Skill 4:** Adds on, using numbers 1–10, with manipulatives	**Skill 5:** Demonstrates counting to 100	**Skill 6:** Has one-to-one correspondence for numbers 1–30	**Skill 7:** Understands combinations (up to 10)	**Skill 8:** Recognizes number groups without counting (2–10)

Kindergarten

Essential Math Skills Class Inventory

Directions: List student names in the first column. When a student reaches proficiency with a skill, record the date in the column under that skill. Update the inventory each week.

	Pre-Kindergarten Skills				Kindergarten Skills				Grade 1 Skills				
Student Name	**Skill 1:** Demonstrates one-to-one correspondence for numbers 1–10, with steps	**Skill 2:** Demonstrates one-to-one correspondence for numbers 1–10, with manipulatives	**Skill 3:** Adds on using numbers 1–10, with steps	**Skill 4:** Adds on using numbers 1–10, with manipulatives	**Skill 5:** Demonstrates counting to 100	**Skill 6:** Has one-to-one correspondence for numbers 1–30	**Skill 7:** Understands combinations (within 10)	**Skill 8:** Recognizes number groups without counting (2–10)	**Skill 9:** Counts objects with accuracy to 100	**Skill 10:** Replicates visual or movement patterns	**Skill 11:** Understands concepts of adding on or taking away (to 30)	**Skill 12:** Adds/subtracts single-digit numbers on paper	**Skill 13:** Shows a group of objects by number (to 100)

Grade 1
Essential Math Skills Class Inventory

Directions: List student names in the first column. When a student reaches proficiency with a skill, record the date in the column under that skill. Update the inventory each week.

	Kindergarten Skills				Grade 1 Skills					Grade 2 Skills					
Student Name	**Skill 5:** Demonstrates counting to 100	**Skill 6:** Has one-to-one correspondence for numbers 1–30	**Skill 7:** Understands combinations (within 10)	**Skill 8:** Recognizes number groups without counting (2–10)	**Skill 9:** Counts objects with accuracy to 100	**Skill 10:** Replicates visual or movement patterns	**Skill 11:** Understands concepts of adding on or taking away (to 30)	**Skill 12:** Adds/subtracts single-digit numbers on paper	**Skill 13:** Shows a group of objects by number (to 100)	**Skill 14:** Quickly recognizes number groups (to 100)	**Skill 15:** Adds/subtracts from a group of objects (to 100)	**Skill 16:** Adds/subtracts two-digit numbers on paper	**Skill 17:** Counts by 2, 3, 4, 5, and 10, using manipulatives	**Skill 18:** Solves written and oral story problems using the correct operations (addition and subtraction)	**Skill 19:** Understands/identifies place value to 1,000

Grade 2

Essential Math Skills Class Inventory

Directions: List student names in the first column. When a student reaches proficiency with a skill, record the date in the column under that skill. Update the inventory each week.

	Grade 1 Skills					Grade 2 Skills						Grade 3 Skills									
Student Name	**Skill 9:** Counts objects with accuracy to 100	**Skill 10:** Replicates visual or movement patterns	**Skill 11:** Understands concepts of adding on or taking away (to 30)	**Skill 12:** Adds/subtracts single-digit numbers on paper	**Skill 13:** Shows a group of objects by number (to 100)	**Skill 14:** Quickly recognizes number groups (to 100)	**Skill 15:** Adds/subtracts from a group of objects (to 100)	**Skill 16:** Adds/subtracts two-digit numbers on paper	**Skill 17:** Counts by 2, 3, 4, 5, and 10, using manipulatives	**Skill 18:** Solves written and oral story problems using the correct operations (addition and subtraction)	**Skill 19:** Understands/ identifies place value to 1,000	**Skill 20:** Reads and writes numbers to 10,000 in words and numerals	**Skill 21:** Uses common units of measurement: length, weight, time, money, and temperature	**Skill 22:** Adds/subtracts three-digit numbers on paper with regrouping	**Skill 23:** Rounds numbers to the nearest 10	**Skill 24:** Rounds numbers to the nearest 100	**Skill 25:** Adds/subtracts two-digit numbers mentally	**Skill 26:** Counts by 5, 6, 7, 8, 9, 10, using manipulatives	**Skill 27:** Uses arrays to visually represent multiplication	**Skill 28:** Recognizes basic fractions	**Skill 29:** Solves written and oral story problems using the correct operation (addition, subtraction, and grouping)

Grade 3

Essential Math Skills Class Inventory

Directions: List student names in the first column. When a student reaches proficiency with a skill, record the date in the column under that skill. Update the inventory each week.

	Grade 2 Skills						Grade 3 Skills									
Student Name	**Skill 14:** Quickly recognizes number groups (to 100)	**Skill 15:** Adds/subtracts from a group of objects (to 100)	**Skill 16:** Adds/subtracts two-digit numbers on paper	**Skill 17:** Counts by 2, 3, 4, 5, and 10 using manipulatives	**Skill 18:** Solves written and oral story problems using the correct operations (addition and subtraction)	**Skill 19:** Understands/ identifies place value to 1,000	**Skill 20:** Reads and writes numbers to 10,000 in words and numerals	**Skill 21:** Uses common units of measurement: length, weight, time, money, and temperature	**Skill 22:**Adds/subtracts three-digit numbers on paper with regrouping	**Skill 23:** Rounds numbers to the nearest ten	**Skill 24:** Rounds numbers to the nearest hundred	**Skill 25:** Adds/subtracts two-digit numbers mentally	**Skill 26:** Counts by 5, 6, 7, 8, 9, 10, using manipulatives	**Skill 27:** Uses arrays to visually represent multiplication	**Skill 28:** Recognizes basic fractions	**Skill 29:** Solves written and oral story problems using the correct operation (addition, subtraction, and grouping)

Proficiency Checklist

Directions: Use this form to record a student's progress toward proficiency in a skill. The student must show proficiency as described in the rubric on at least three different occasions, using three different instructional activities to be considered proficient. Noting proficiency is assurance that the student has learned this skill such that it will not regress and be lost at some time in the future.

Student: ______________________________ Date: ________________

Activity	Date	Evaluation (Emerging/Intervention, Developing, or Proficient)	Notes

Pre-Kindergarten
Skill Progression Rubric

Student: ________________________________ Date: ________________

Skill	Emerging	Developing	Proficient
Skill 1: Demonstrates one-to-one correspondence for numbers 1–10, with steps	Not yet able to count steps in sequence	Able to count 2 to 9 steps in sequence but is sometimes inconsistent	Counts 10 or more steps in sequence
Skill 2: Demonstrates one-to-one correspondence for numbers 1–10, with manipulatives	Not yet counting objects with one-to-one correspondence	Counts objects with accuracy to 3	Shows one-to-one correspondence when counting 10 or more objects
Skill 3: Adds on, using numbers 1–10, with steps	Unable to add on numbers using steps on a number line without recounting (1–10)	Inconsistently adds on numbers using steps on a number line without recounting (1–10)	Adds on numbers using steps on a number line without recounting (1–10)
Skill 4: Adds on, using numbers 1–10, with manipulatives	Unable to add on numbers without recounting (1–10)	Inconsistently adds on from a set of objects without recounting (1–10)	Adds on from a set of objects without recounting (1-10)

Kindergarten

Skill Progression Rubric

Student: ______________________________ Date: ________________

Skill	Intervention	Developing	Proficient
Skill 5: Demonstrates counting to 100	Counts to a number less than 30 with accuracy	Counts to 31–99 with accuracy	Counts to 100 with accuracy
Skill 6: Has one-to-one correspondence for numbers 1–30	Can count fewer than 10 objects with one-to-one correspondence	Shows one-to-one correspondence when counting 10 to 29 objects	Shows one-to-one correspondence when counting 30 or more objects
Skill 7: Understands combination (within 10)	Does not consistently understand how to add on or take away 1–3 objects without recounting	Consistently adds on or takes away 1-3 objects without recounting	Adds on or takes away from a set of objects without recounting (1–10)
Skill 8: Recognizes number groups without counting (2–10)	Recognizes number groups of 1 or 2 without counting individual objects	Recognizes number groups of 3 to 5 without counting individual objects	Recognizes number groups up to 10 without counting individual objects

Grade 1

Skill Progression Rubric

Student: ______________________ Date: ______________

Skill	Intervention	Developing	Proficient
Skill 9: Counts objects with accuracy to 100	Counts fewer than 20 objects with accuracy	Sometimes counts objects with accuracy (20 to 100)	Consistently counts objects with accuracy (to 100)
Skill 10: Replicates visual or movement patterns	Has difficulty replicating a two-step visual pattern (e.g., square, circle; or red, blue) or a two-step movement pattern (e.g., clap hands, step forward)	Can sometimes replicate a two- or three-step visual or movement pattern	Can consistently replicate three- and four-step visual and movement patterns
Skill 11: Understands concepts of adding on and taking away (to 30), with manipulatives	Unable to add on or take away numbers from a group (within 10)	Uses manipulatives to add on or take away from a group, but must recount to find the total (within 30)	Using manipulatives, can add on or take away numbers and name the resulting number (within 30)
Skill 12: Adds/subtracts single-digit numbers on paper	Needs assistance to add two single-digit numbers on paper	Adds two single-digit numbers on paper independently and with partial accuracy	Adds or subtracts two single-digit numbers on paper independently and accurately
Skill 13: Shows a group of objects by number (to 100)	Shows a group of objects of fewer than 25, using manipulatives	Shows a group of objects to 50, using manipulatives	Shows a group of objects to 100, using manipulatives

Grade 2

Skill Progression Rubric

Student: ______________________ Date: ______________

Skill	Intervention	Developing	Proficient
Skill 14: Quickly recognizes groups of objects (to 100)	Using manipulatives, recognizes number groups of fewer than 25	Using manipulatives, recognizes number groups of 25–75	Using manipulatives, quickly recognizes number groups to 100
Skill 15: Adds to/ subtracts from a group of objects (within 100)	Needs assistance to add to a group of objects and recognize the sum	Accurately adds to a group of objects but needs assistance with subtraction	Accurately adds to and subtracts from a group of objects within 100
Skill 16: Adds/subtracts two-digit numbers on paper	Needs assistance to add two-digit numbers on paper	Accurately adds two-digit numbers on paper but needs assistance with subtraction	Accurately adds and subtracts two-digit numbers on paper
Skill 17: Counts by 2, 3, 4, 5, and 10, using manipulatives	Unable to skip-count using manipulatives	Consistently counts by 2 and 5, using manipulatives	Consistently counts by 2, 3, 4, 5, and 10, using manipulatives
Skill 18: Solves written and oral story problems using the correct operations *(addition and subtraction)*	Unable to solve written and/or oral story problems with guidance	Able to solve written and oral story problems with guidance	Consistently able to solve written and oral story problems by developing a plan, solving problems using correct operations, and evaluating the solution
Skill 19: Understands/ identifies place value to 1,000	Does not yet understand place value for the ones, tens, hundreds, or thousands place	Consistently identifies place value for the ones and tens place. Sometimes identifies place value for hundreds and/or thousands place	Consistently understands and identifies place value for the ones, tens, hundreds, and thousands places

Grade 3
Skill Progression Rubric

Student: ______________________________ Date: ________________

Skill	Intervention	Developing	Proficient
Skill 20: Reads and writes numbers to 10,000 in words and numerals	Counts, reads, and/or writes numbers to 100	Reads and/or writes numbers to 1,000	Reads and writes numbers to 10,000
Skill 21: Uses common units of measurement: length, weight, time, money, and temperature	Does not yet use units of measurement accurately	Able to use some units of measurement with accuracy	Consistently able to use all common units of measurement accurately
Skill 22: Adds/subtracts three-digit numbers on paper with regrouping	Unable to add or subtract three-digit numbers with regrouping	Able to accurately add and subtract three-digit numbers with guidance or use of manipulatives	Able to accurately add and subtract three-digit numbers on paper
Skill 23: Rounds numbers to the nearest ten	Unable to round numbers to the nearest ten	Rounds numbers to the nearest ten sometimes but not yet consistently	Consistently rounds numbers to the nearest ten
Skill 24: Rounds numbers to the nearest hundred	Unable to round numbers to the nearest hundred	Rounds numbers to the nearest hundred sometimes but not yet consistently	Consistently rounds numbers to the nearest hundred
Skill 25: Adds/subtracts two-digit numbers mentally	Unable to mentally add or subtract two-digit numbers	Sometimes able to demonstrate ability to mentally add and/or subtract two-digit numbers	Consistently demonstrates ability to mentally add and subtract two-digit numbers
Skill 26: Counts by 5, 6, 7, 8, 9, and 10, using manipulatives	Unable to skip-count using manipulatives	Skip-counts by 5 and 10 but struggles with 6, 7, 8, or 9, using manipulatives	Consistently counts by 5, 6, 7, 8, 9, and 10, using manipulatives
Skill 27: Uses arrays to visually represent multiplication	Uses arrays to visually represent multiplication problems with a factor of 2	Uses arrays to visually represent multiplication problems with factors through 6	Uses arrays to visually represent multiplication for problems with factors through 12
Skill 28: Recognizes basic fractions	Unable to identify $\frac{1}{2}$, $\frac{1}{3}$, and $\frac{1}{4}$ in picture form with automaticity	Able to identify basic fractions in pictures and draw basic fractions	Able to identify, draw, and write basic fractions
Skill 29: Solves written and oral story problems using the correct operation *(addition, subtraction, and grouping)*	Able to solve written and/or oral story problems if given plan and correct operation	Able to solve written and oral story problems with guidance in developing plan	Consistently able to solve written and oral story problems by developing a plan, using correct operations, and evaluating the solution

Name: ____________________ Date: ____________

Here's What I Think

Directions: Show your thinking about how you found your solution. Use pictures, numbers, or words.

Skill: ______________________________

Activity: ______________________________

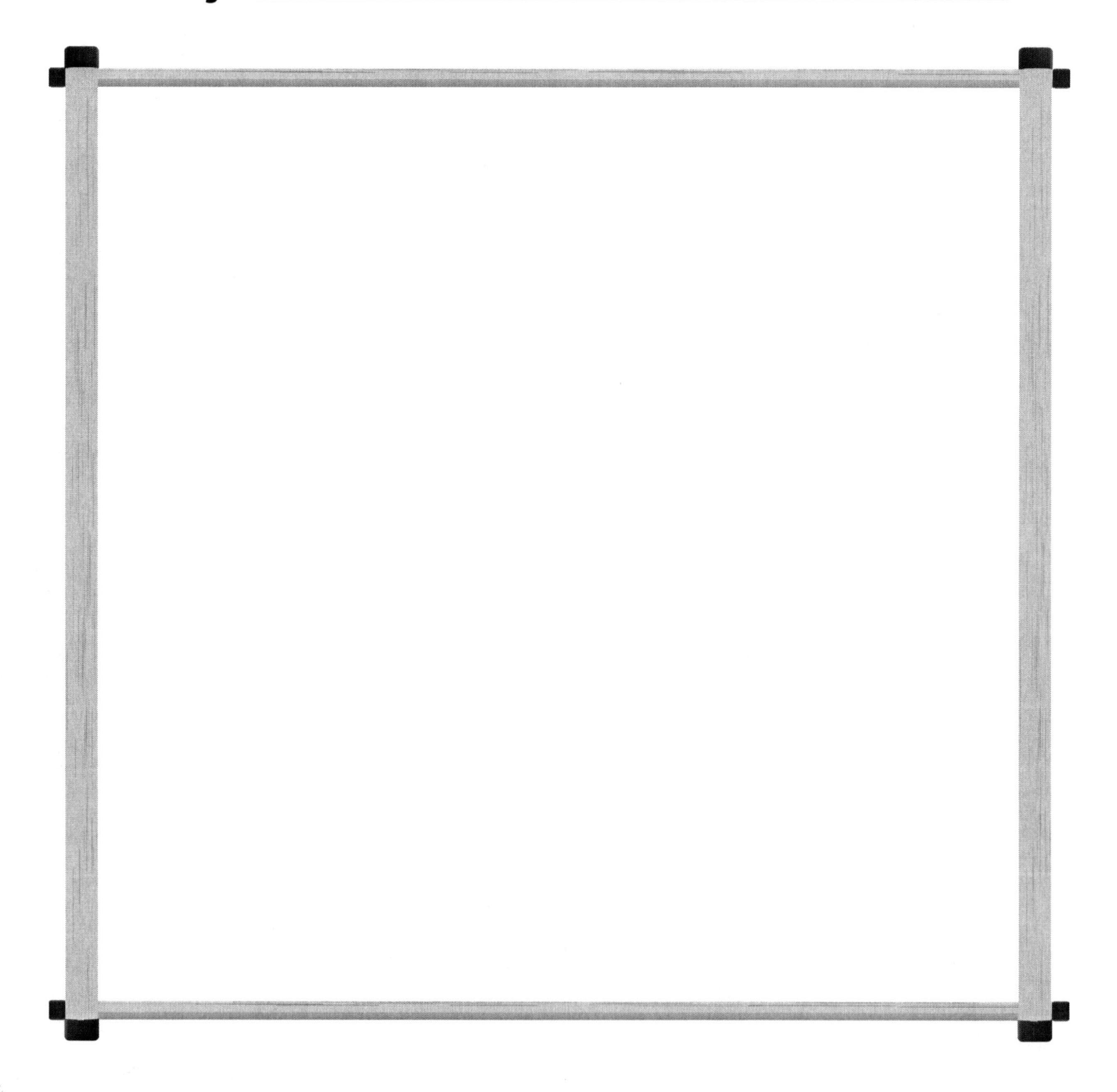

Name: ____________________ Date: ____________

Place-Value Chart

Directions: Look at the numeral in the left column. Put the correct digit in the ten thousands, thousands, hundreds, tens, and ones columns.

Numeral	Ten thousands	Thousands	Hundreds	Tens	Ones	Number value in words

Name: ______________________ Date: ______________

Hundred Chart

1	2	3	4	5	6	7	8	9	10
11	12	13	14	15	16	17	18	19	20
21	22	23	24	25	26	27	28	29	30
31	32	33	34	35	36	37	38	39	40
41	42	43	44	45	46	47	48	49	50
51	52	53	54	55	56	57	58	59	60
61	62	63	64	65	66	67	68	69	70
71	72	73	74	75	76	77	78	79	80
81	82	83	84	85	86	87	88	89	90
91	92	93	94	95	96	97	98	99	100

Name: ______________________ Date: ____________

Blank Hundred Chart

Name: ______________________________ Date: ________________

Fifty Chart

1	2	3	4	5	6	7	8	9	10
11	12	13	14	15	16	17	18	19	20
21	22	23	24	25	26	27	28	29	30
31	32	33	34	35	36	37	38	39	40
41	42	43	44	45	46	47	48	49	50

Name: ______________________________ Date: ________________

Rounds Up or Down Hundred Chart

Directions: Put your finger on the number given by your teacher. Then, round it up or down to the nearest ten.

Nearest Ten	← Round Down				→ Round Up					Nearest Ten
0	1	2	3	4	5	6	7	8	9	10
10	11	12	13	14	15	16	17	18	19	20
20	21	22	23	24	25	26	27	28	29	30
30	31	32	33	34	35	36	37	38	39	40
40	41	42	43	44	45	46	47	48	49	50
50	51	52	53	54	55	56	57	58	59	60
60	61	62	63	64	65	66	67	68	69	70
70	71	72	73	74	75	76	77	78	79	80
80	81	82	83	84	85	86	87	88	89	90
90	91	92	93	94	95	96	97	98	99	100

Name: ______________________________ Date: ______________

Round to the Nearest Ten

Directions: Write the number in the left column. Then, write each digit in the appropriate column. Round each number to the nearest ten.

Numeral	Thousands	Hundreds	Tens	Ones	Rounded Value
544		5	4	4	540
4,687	4	6	8	7	4,690

Name: ______________________________ Date: ______________

Round to the Nearest Hundred

Directions: Write the number in the left column. Then, write each digit in the appropriate column. Round each number to the nearest hundred.

Numeral	Ten Thousands	Thousands	Hundreds	Tens	Ones	Rounded Value
6,544		6	5	4	4	6,500
24,687	2	4	6	8	7	24,700

Name: ______________________________ Date: ______________

Mental Math Bingo

Teacher Directions: Pick one number between 25 and 100. Read the directions below to students, and then tell them the starting number.

Directions: Find the quickest and easiest solutions that will allow you to answer five questions in a row. Your correct answers can be vertical, horizontal, or diagonal. All computation must be done in your head. Do not use paper, an abacus, or a calculator to help you find the answer. Write your answers in the appropriate box, and raise your hand when you have bingo!

Beginning number: __________

Double it	Add 35	Subtract 12	Add 17	Add 100
Subtract 14	Subtract 15	Subtract 3	Add 83	Add 99
Add 17, and then subtract 5	Add 30	Half	Subtract 11	Double it, and subtract 10
Add 35	Subtract 17	Add 58	Subtract 15	Subtract 23
Subtract 8	Add 25	Add 49	Double it	Triple it

Name: ______________________________________ Date: __________________

Fill in the Fraction

Directions: Shade ____________ of each figure or set.

1. 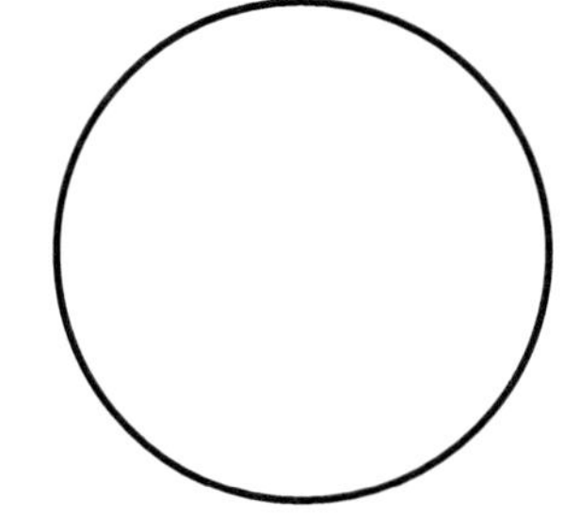

2. 

3. 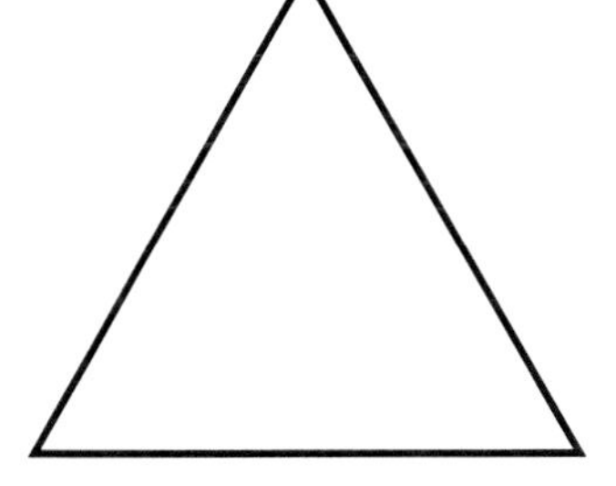

4. 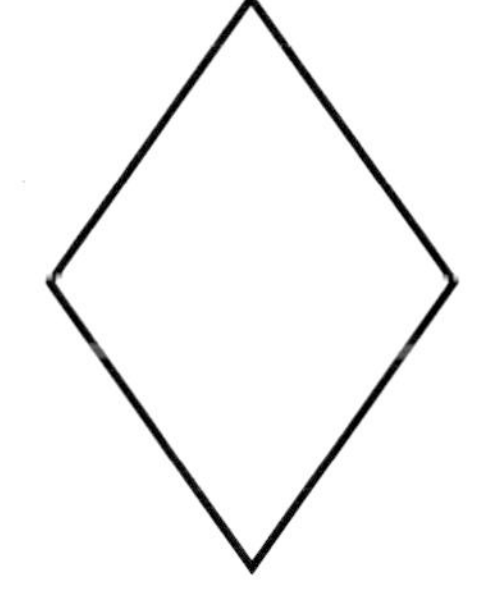

5. 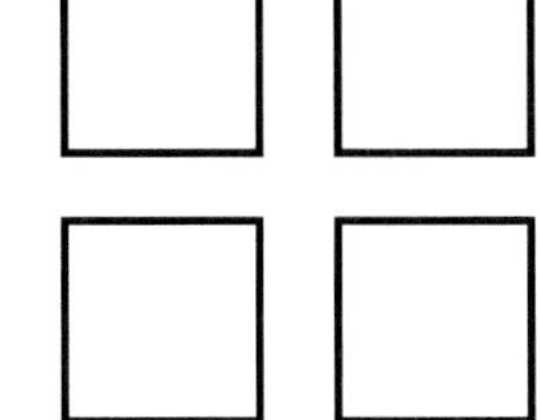

6. 

7. 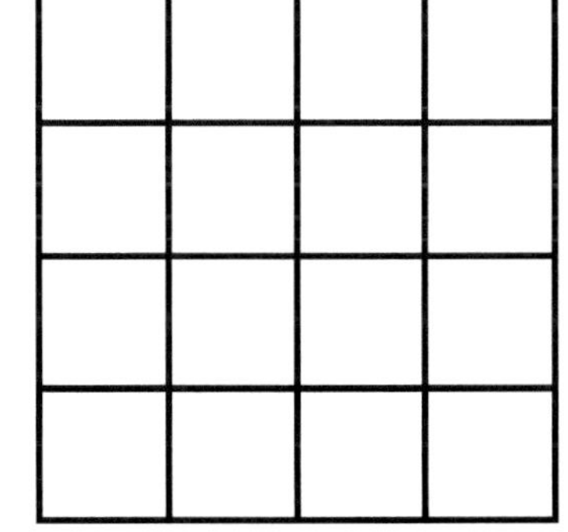

8.

Name: ______________________________ Date: ______________

Draw Fractions

Directions: Draw a picture to show each fraction.

1. $\frac{1}{2}$	2. four-fifths	3. two-fifths
4. $\frac{3}{6}$	5. seven-eighths	6. three-fourths
7. $\frac{8}{10}$	8. six-sevenths	9. $\frac{1}{3}$
10. one-half	11. $\frac{2}{3}$	12. four-sixths

Name: ______________________________ Date: ______________

Matching Fractional Values

Teacher Directions: In the left column, draw pictures or use words to represent fractional values. In the right column, show those fractions in numeral form, mixing up the order in which the fractions are given.

Directions: Draw a line to match each picture or word in the left column with the fraction it represents in the right column.

Name: ______________________ Date: ______________

What Fraction Is Shaded?

Directions: Identify the fraction of each figure that is shaded. Write the fraction in simplest form, if possible.

1. 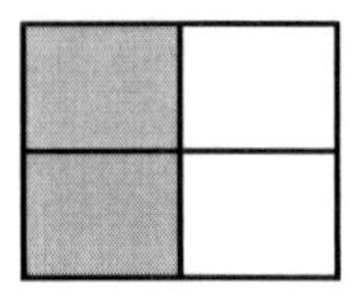

____

2. 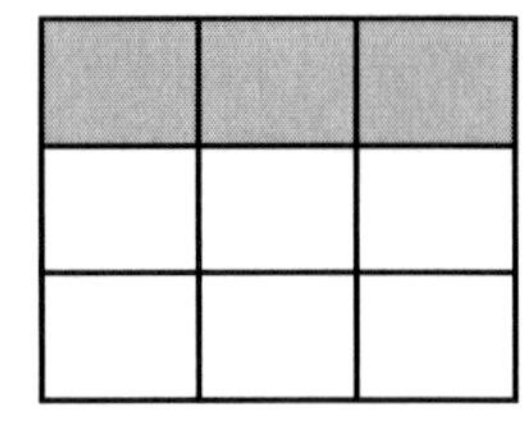

____

3. 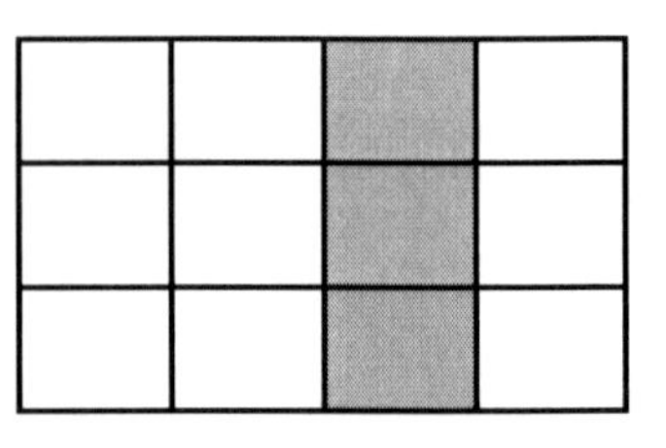

____

4. 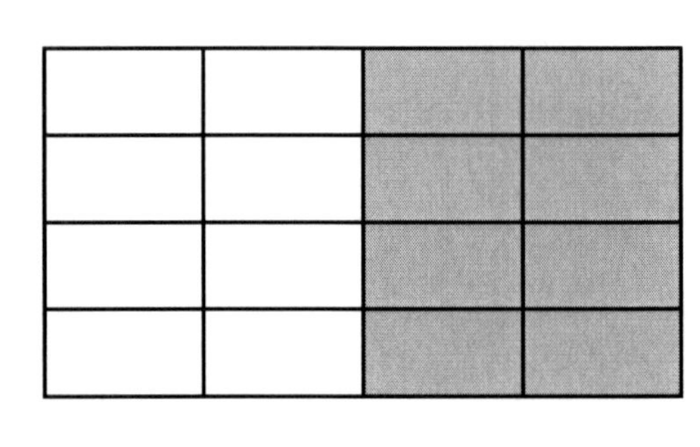

____

5. 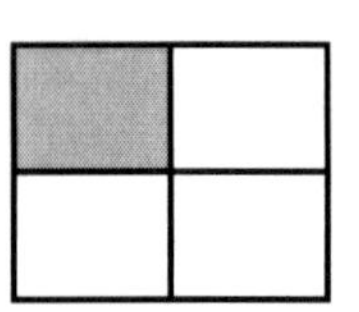

____

6. 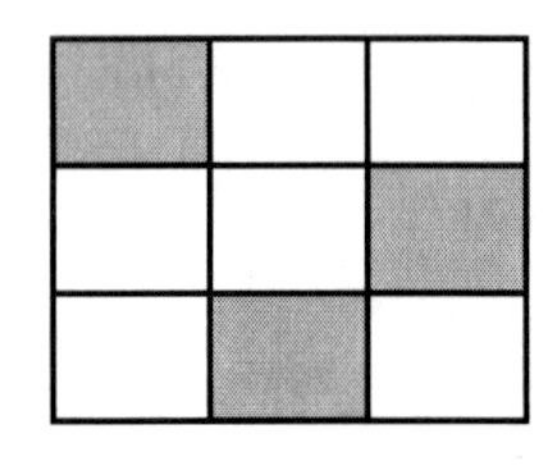

____

7. 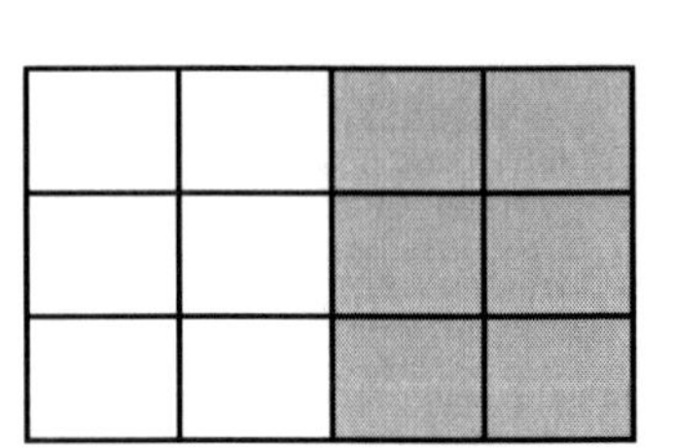

____

8. 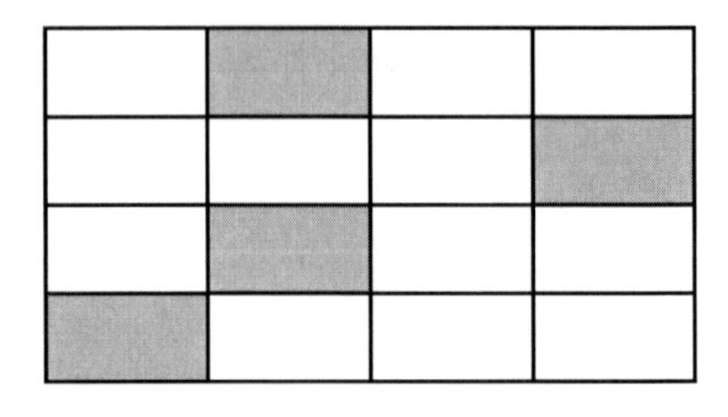

____

Name: ______________________________ Date: ______________

What Fraction Is Shaded? 2

Teacher Directions: Shade a portion of each figure before distributing the activity sheet to students.

Directions: Identify the fraction of each figure that is shaded. Write the fraction in simplest form, if possible.

1. 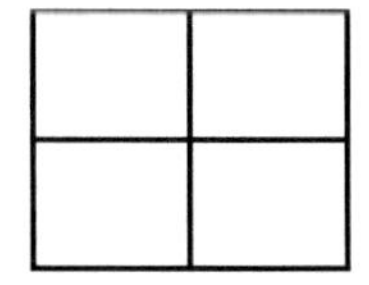

____

2. 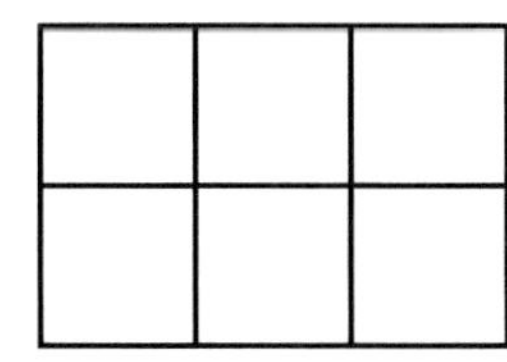

____

3. 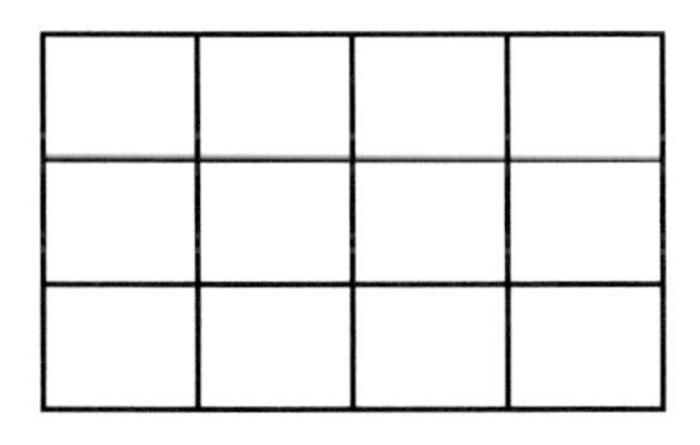

____

4. 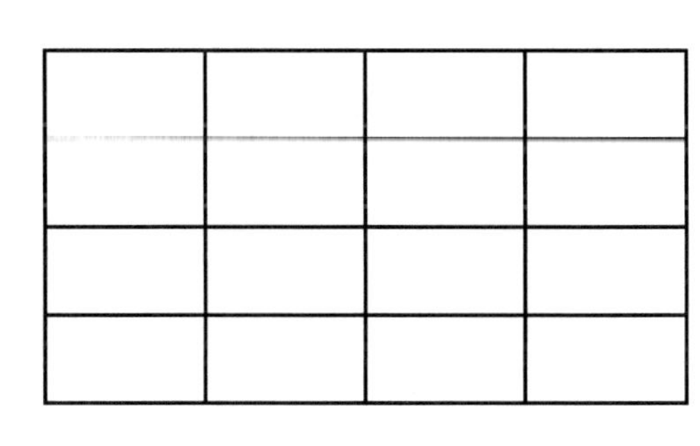

____

5. 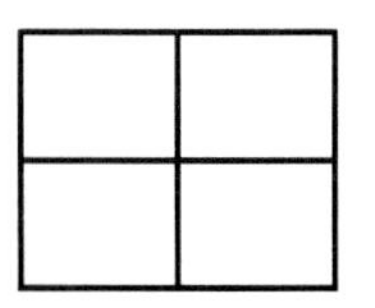

____

6. 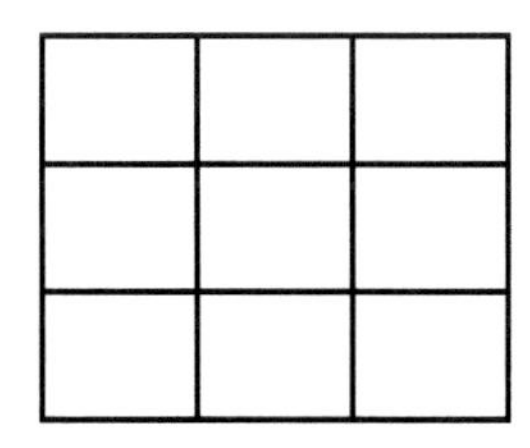

____

7. 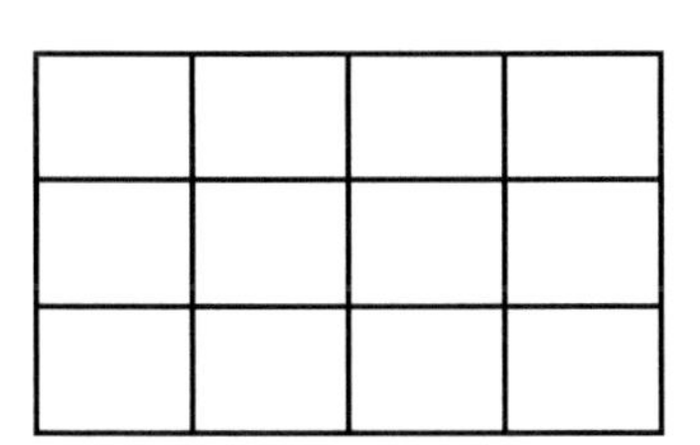

____

8. 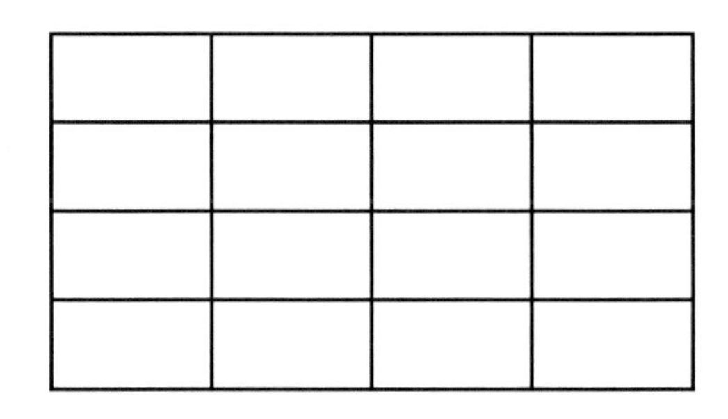

____

Name: ______________________________ Date: ______________

Dollars, Dimes, and Pennies Array

Directions: Draw the dollars, dimes, and pennies for each step in a money problem.

Dollars	Dimes	Pennies	Find the Sum
			+
			Total

Dollars	Dimes	Pennies	Find the Sum
			+
			Total

Name: ______________________________ Date: ______________

Number Dollars, Dimes, and Pennies Array

Directions: Write the money value in the right column, and draw the correct number of dollars, dimes, and pennies in that row.

Variation: *Write the money value in the right column, and write the numeral for dollars, dimes, and pennies in that row.*

Dollars	Dimes	Pennies	Total Value

Name: ______________________________ Date: ______________

Problem-Solving Bingo

Teacher Directions: Write the answers to each of 25 short story problems in the squares. Create several versions by rearranging the answers on the cards.

Directions: Mark the correct answer to each story problem. Raise your hand when you match 5 problems to their solutions (vertically, horizontally, or diagonally). You have bingo!

Family Letter

Dear Family,

I am working hard to help your child develop the mathematics skills that will serve him or her in our information- and technology-rich world. We are working toward mastery of identified grade-level skills that are essential for long-term success in math, and if I notice a skill that needs a little bit of extra practice at home, I will contact you and show you how you can support your child's math development.

In the meantime, there are many ways to help your child have the experiences that help build an awareness of math concepts and purposes. Please look for ways to offer your child experiences outside school that promote the following:

- **Awareness of Distance:** Activities might include climbing, reaching, going for walks, crawling, riding bikes, or throwing balls.
- **Awareness of Weight:** Activities might include putting away groceries, pouring water into a cup or bucket, comparing plastic toys to wooden or metal ones, carrying rocks, putting away toys, or climbing.
- **Awareness of Patterns:** Activities might include setting the table, playing with multicolored blocks, planting a garden, climbing stairs, coloring, cutting, drawing, crafts, playing hopscotch, or skipping.
- **Awareness of Frequency:** Activities might include taking turns, listening to music, playing a musical instrument, or counting objects.
- **Awareness of Time:** Activities might include waiting for dinner, waiting for a parent to finish a phone call, monitoring the position of the sun, watching the seasons change, counting the days until a holiday, or reading clocks.
- **Awareness of Equality and Equations:** Activities might include playing on a seesaw, playing with sand or water, building with blocks, balancing on a balance board or a beam, trading with a sibling, or dividing food into portions and serving it.

Students will be able to work with math abstractly only after they have had sufficient learning experiences with real objects in the physical world. By offering your child many of these learning opportunities, you will help him or her to be successful in math both at school and in life.

Sincerely,

____

Your Child's Teacher

Math Vocabulary

Sometimes, we assume that children come to school already understanding the words that help us think about mathematics and numbers. But it is often necessary to specifically include instruction on the meaning of these vocabulary terms. Consider posting these words on a Math Word Wall, incorporating them into conversations with students, and including vocabulary activities as part of your preparation for an upcoming math lesson.

- Above
- Add
- Addition
- After
- Alike
- Always
- Array
- As many as
- Away
- Backward
- Before
- Beginning
- Behind
- Below
- Between
- Bottom
- Calendar
- Cent
- Center
- Centimeter
- Circle
- Clock
- Coin
- Compare
- Corner
- Count
- Cup
- Day
- Difference
- Different
- Digit
- Dime
- Dollar
- Down
- End
- Equal
- Every
- Farthest
- Few
- Fewer
- Fewest
- First
- Foot
- Forward
- Fraction
- Front
- Half
- Half-dollar
- Half-hour
- Hour
- Hour hand
- Hundred
- Inch
- Kilometer
- Last
- Least
- Left
- Length
- Less
- Less than
- Liter
- Long
- Longer
- Longest
- Measure
- Meter
- Mile
- Minus
- Minute
- Money
- Month
- More
- More than
- Most
- Near
- Never
- Next
- Nickel
- Number
- Number line
- Number sentence
- Numeral
- One
- Operation
- Opposite
- Order
- Other
- Over
- Pair
- Part
- Pattern
- Penny
- Pint
- Place value
- Plus
- Pound
- Quart
- Quarter
- Rectangle
- Right
- Row
- Rule
- Ruler
- Same
- Season
- Second
- Second hand
- Separate
- Sequence
- Several
- Shape
- Short
- Shorter
- Shortest
- Side
- Sides
- Size
- Skip
- Small
- Smaller
- Smallest
- Solution
- Solve
- Some
- Sort
- Square
- Starting
- Subtract
- Subtraction
- Sum
- Temperature
- Ten
- Thermometer
- Third
- Thousand
- Three-dimensional
- Through
- Top
- Total
- Triangle
- Two-dimensional
- Under
- Up
- Week
- Whole
- Whole number
- Wide
- Wider
- Widest
- Width
- Yard
- Year
- Zero

References Cited

Hernandez, Donald J. 2011. "Double Jeopardy: How Third-Grade Reading Skills and Poverty Influence High School Graduation." Baltimore, MD: The Annie E. Casey Foundation.

National Assessment of Educational Progress. 2009. "The Nation's Report Card." http://nationsreportcard.gov/math_2009/gr4_national.asp.

National Assessment of Educational Progress. 2011. "The Nation's Report Card." http://nces.ed.gov/nationsreportcard/mathematics.

National Mathematics Advisory Panel. 2008. "Foundations for Success: The Final Report of the National Mathematics Advisory Panel." Washington, DC: U.S. Department of Education.

Shore, Rima. 2009a. "PreK–3rd: What Is the Price Tag?" New York: Foundation for Child Development. http://fcd-us.org/resources/preK-3rd-what-price-tag.

Shore, Rima. 2009b. "The Case for Investing in PreK–3rd Education: Challenging Myths about School Reform." New York: Foundation for Child Development. http://fcd-us.org/sites/default/files/TheCaseForInvesting-ChallengingMyths.pdf.

The Annie E. Casey Foundation. 2010. "Early Warning! Why Reading by the End of Third Grade Matters." Baltimore, MD: The Annie E. Casey Foundation.

Contents of the Digital Resource CD

Student Resources

Page	Title	Filename
148	Here's What I Think	whatthink.pdf
149	Place–Value Chart	placevaluechart.pdf
150	Hundred Chart	hundredchart.pdf
151	Blank Hundred Chart	blankhundred.pdf
152	Fifty Chart	fiftychart.pdf
153	Rounds Up or Down Hundred Chart	roundsup.pdf
154	Round to the Nearest Ten	roundnearestten.pdf
155	Round to the Nearest Hundred	roundnearesthundred.pdf
156	Mental Math Bingo	mentalmath.pdf
157	Fill in the Fraction	fillinfraction.pdf
158	Draw Fractions	drawfractions.pdf
159	Matching Fractional Values	matchingfractional.pdf
160	What Fraction Is Shaded?	whatfraction.pdf
161	What Fraction Is Shaded? 2	whatfraction2.pdf
162	Dollars, Dimes, and Pennies Array	dollarsarray.pdf
163	Number Dollars, Dimes, and Pennies Array	numbersarray.pdf
164	Problem-Solving Bingo	probsolvingbingo.pdf
N/A	Ten Chart	tenchart.pdf
N/A	Twenty Chart	twentychart.pdf
N/A	"Make 1" Cards	make1.pdf
N/A	Thirty Chart	thirtychart.pdf
N/A	Picture Cards	picturecards.pdf
N/A	Place-Value Cards	placevaluecards.pdf
N/A	Skill 18 Story Problems	skill18problems.pdf
N/A	Greatest Sum Cards	greatestsum.pdf
N/A	Greatest Sum Board	sumboard.pdf
N/A	Spinner	spinner.pdf
N/A	Skill 29 Story Problems	skill29problems.pdf
N/A	Problem-Solving Bingo Story Problems	bingoproblems.pdf

Teacher Resources

Page	Title	Filename
14–18	Correlation to Standards	standards.pdf
135–136	The Pre-K to Grade 3 Essential Math Skills Individual Inventory	individualinventory.pdf; individualinventory.xls
137	Pre-Kindergarten Essential Math Skills Class Inventory	prekinventory.pdf; prekinventory.xls
138	Kindergarten Essential Math Skills Class Inventory	kinderinventory.pdf; kinderinventory.xls
139	Grade 1 Essential Math Skills Class Inventory	grade1inventory.pdf; grade1inventory.xls
140	Grade 2 Essential Math Skills Class Inventory	grade2inventory.pdf; grade2inventory.xls
141	Grade 3 Essential Math Skills Class Inventory	grade3inventory.pdf; grade3inventory.xls
142	Proficiency Checklist	profchecklist.pdf
143	Pre-Kindergarten Skill Progression Rubric	prekskillrubric.pdf
144	Kindergarten Skill Progression Rubric	kinderskillrubric.pdf
145	Grade 1 Skill Progression Rubric	grade1skillrubric.pdf
146	Grade 2 Skill Progression Rubric	grade2skillrubric.pdf
147	Grade 3 Skill Progression Rubric	grade3skillrubric.pdf
165	Family Letter	familyletter.docx; familyletter.pdf
166	Math Vocabulary	mathvocab.docx; mathvocab.pdf